Questing

Questing

A Guide to Creating
Community Treasure Hunts

Delia Clark & Steven Glazer
Foreword by David Sobel

University Press of New England
Hanover and London

Published by University Press of New England,
One Court Street, Lebanon, NH 03766
www.upne.com
© 2004 by Antioch New England Institute and Vital Communities
Printed in the United States of America
5 4 3 2 1

Book design by Dean Bornstein

Library of Congress Cataloging-in-Publication Data
Clark, Delia.
Questing : a guide to creating community treasure hunts / Delia Clark and
Steven Glazer ; foreword by David Sobel.
    p.   cm.
Includes bibliographical references and index.
ISBN 1–58465–334–5 (cloth : alk. paper)
    1. Place-based education—United States.   I. Glazer, Steven.   II. Title.
LC239.C553 2004
370.11'5—dc22   2004005369

The author and publisher gratefully acknowledge the following:

Excerpts from *Pilgrim at Tinker Creek,* by Annie Dillard, appear with the permission of Annie Dillard.

"How to See Deer," copyright © 1975 by Phillip Booth, from *Relations: New
and Selected Poems* by Phillip Booth. Used by permission of Viking Penguin, a
division of Penguin Group (USA).

Arthur Sze, "The Network," from *The Redshifting Web: Poems 1970–1998.*
Copyright © 1998 by Arthur Sze. Reprinted with permission of Copper
Canyon Press, P.O. Box 271, Port Townsend, WA 98368-0271.

For Susan Clark
and Chagdud Tulku Rinpoche,
with gratitude and love

The world is all clues, and there is no end
to their subtlety and delicacy. The signs that
reveal are always there. One has only to learn
the art of reading them.

—Paul Shepard, *Nature and Madness*

CONTENTS

FOREWORD

David Sobel

Lean close. Let me tell you a secret. On the surface, this looks like a charming little book about treasure hunts, good education, building community. All that apple pie and motherhood stuff. But underneath it's really much more—it's a Buddhist tract on right livelihood, a cabalistic guide to changing your life, a witch's spellbook. It's about a different way of seeing. A few Novembers ago, at the end of an afternoon of burning brush, my ten-year-old son returned from a far-ranging woods exploration. Under the silvery light of a crescent moon, standing ankle deep in fresh snow, he said, "I'm a good explorer because I really look at all the details, all the little places you can go, all the crannies you can find. I don't just look at it and go, I spend a lot of time on it, make forts and stuff and traps." This is a book about all the crannies you can find, about how to help others poke into those crannies, and about how going on and creating Quests can become an organizing metaphor for your life. Let me share a personal example of how Questing has watercolored my life.

I had a perfect Quest-filled day in Cornwall and Devon, England, a few years ago. My twelve-year-old daughter and I were on three Quests. We were searching for King Arthur's birthplace, for the perfect cream tea, and for the inner circle of "Letterboxing" aficionados (Letterboxing is the British ancestor of Questing).

For a few years prior to our trip, my daughter had immersed herself in Arthurian legend, so our goal was to visit Tintagel, the castle where Arthur was supposed to have been born to Uther and Igraine. The weather was classically British—dreary, misty, rainy, and windswept. We had to weave along miles of hedge-hidden country lanes, and confusion arose at every road junction. When we made it to the village adjacent to the castle ruins, we took refuge in the King Arthur Bookstore and tried not to dampen the book jacket with drops from our raincoats as we reread the opening passage of *The Mists of Avalon.* Here, perched on a rocky headland, Igraine stares out over the windswept Atlantic on a raw June day, searching the waters for a sign. Our goal was to find the actual location of Igraine's lookout.

Struggling to keep the wind from turning our umbrella inside out, we climbed up the stone steps, through the medieval portico, and into the castle ruins. With the text echoing in our minds, we scrambled off the designated paths to test rocky perches that fit the bill. We had to be able to look north, be directly above a jagged cliff, and be somewhat away from the castle proper. At one outcrop, I held Tara's belt as she leaned out over a declivity to peer down at the waves crashing below. Could this be the place? We rushed past intermingled ghosts and tourists as we considered all our options. Nope, not exposed enough. Perhaps, but not a great view over the ocean. We're not sure we ever found exactly the right location, or if Marion Zimmer Bradley actually had a specific location in mind when she wrote that opening passage, but what fun we had. As we tucked into our hot lentil soup, snuggled up in an empty pub in the village, we felt valiant that we had braved the elements and at least gotten close.

We then headed north along the coast to Bocastle, a narrow harbored port, in search of that eminently decadent tradition, a Devonshire Cream Tea. In between squalls, we dodged from inn to pub to tea room, checking out our options, when, all the sudden, we stepped into a fairy tale. Chimes jangled as we swung open the wooden door with iron brackets. The scraggly-haired, pointy-bearded, pale-blue-eyed proprietor smiled knowingly as we entered. Patchouli-scented air drew us into a grotto of witch lore, tarot cards, spell books, black-and-white marble chess sets, celtic charms, multifaceted crystals, animal skulls, celtic knots. Harp music emanated from somewhere and merged with the tinkle of running water in a crafted stone garden. Whispering was mandatory. Merlin's spirit lurked in the shadows.

Dazed and dazzled, we drifted into the adjacent tearoom. The yellow porcelain pot of English Breakfast soon arrived, hidden inside a rose cozy. On the table, the tiny pewter candle lamp with a fringed and beaded lampshade provided our private pool of light. Ivy framed the view out over the harbor as we mounded clotted cream and chunky strawberry jam onto still-warm almond scones. It was the pot of gold at the end of the rainbow, the tupperware container at the end of the Quest, our own little holy grail. We had enjoyed other cream teas, but this one managed the perfect blend of mystery and mistiness.

Back across Devon, we skirted the northern fringe of Dartmoor on narrow winding moor lanes with hedges so high we could barely see the sky. The rain abated, Tara napped, and I navigated us to the Dolphin Inn in the pleasant market town of Bovey Tracey. We wove our way past the pub bar, skirted the main dining room, and found the alcove reserved for the Wednesday night gathering of Letterboxers (that's big "L," not small "l," Letterboxers.) This isn't your run-of-the-mill gathering of boy scout leaders and recreational moor walkers. Instead, we had gained access to the inner circle, the power behind the throne, the national security council of Dartmoor lore. And since we'd called ahead, we were discreetly ushered to a table in a darkened corner: Godfrey's table, the lair of the grandmaster himself. Godfrey had been the CEO of a major distillery, but he had chosen early retirement so he could dedicate himself to Letterboxing. Over fish and chips, we interviewed Godfrey about the early years, the fragile peace between Letterboxers and the national park service, the psychology of Letterboxing. Our conversation, however, soon got eclipsed by the evening's featured activity: a set of densely abstruse Letterboxing challenges.

Now, most Letterboxing bears some resemblance to the Questing in this book. Some Letterbox clues are just latitude and longitude coordinates with a rhymed couplet to help you find the box once you're gotten yourself close to the designated location. Other Letterboxes are located by riddles that require you to pore over the classic *Crossing's Guide to Dartmoor*. This 800-page tome is the definitive guide to the thousands of years of history of one of England's best-known eerie highlands. Also popular are sets of Letterboxes organized in a tour as fundraisers for community groups. You send in your two pounds and receive a guided treasure hunt pamphlet leading you to a dozen or so boxes.

But what was going on that evening at the Dolphin was a Quest of a different order. The Letterboxers gathered there that night knew the wide, tumultuous expanse of Dartmoor like the backs of their pale hands. They'd probably all found more than one thousand boxes in their days and had rambled over endless hills and dales. Furthermore, they'd rambled as extensively through the pages of *Crossing's Guide* and many of the other natural and human histories of Dartmoor. Finally, they were all crossword, anagram, and riddle solvers of the highest order. As one woman described to me that evening, "We're Brits. We get very intense about daft things."

Circulating through the crowd was a set of one-line riddles indicating the location of about fifteen recently hidden Letterboxes. The riddles were ridiculously sparse—"The pale sun o'er the rill spangles the robin's breast in the early evening." Within minutes, someone would say, "Oh, that one's easy. I know where that is." By weaving together their geographic, historical, and literary expertise, they could chat for a few minutes and find these boxes in their minds! They'd been out there in the wind-driven drizzle and up late at night committing *Crossing's Guide* to memory so often that they could visualize the location of the boxes. They were solving the *New York Times* Sunday crossword puzzle sprawled out across hundreds of square miles, and they weren't using pens! Now that, I mused to myself, is truly a sense of place. During the chatter, Godfrey slipped out early, and when we went to pay, we found he had surreptitiously picked up our tab. Genteel *and* smart, these Brits.

And so, I invite you to get intense about daft things. But daft only to the untutored mind. The process of making Quests, with or for your students or coworkers or community members or visitors, will inevitably draw you down into the fascinating complexity of local history, ecology, and culture. You'll find out when the first settlers arrived and how their values shaped the character of your community, or you'll discover why blackburnian warblers only show up in one thicket in your town and why it's important for the conservation commission to buy that piece of land. The story of the three brothers from your neighborhood who died at Antietem will help you understand why townspeople erected the Civil War monument outside City Hall. And you might find that people who go on your Quests are more likely to want to run for the school board or help organize the church supper, too.

Of course, you'll also bring joy into people's lives. My kids and I make an annual fall pilgrimage to the John Hay Estate, on the shores of Lake Sunapee in Newbury, New Hampshire. John Hay was Theodore Roosevelt's secretary of state, and his summer "cottage" and grounds are now a little-known national wildlife refuge. The hidden garden, with its winding paths, sequestered pools, statuary-filled grottoes, and massive rhododendrons, is one of our favorite hide-and-seek landscapes. As we emerged with grass-stained knees and twigs in our hair one twilight last October, we came upon a young couple with two toddlers reaching into the recesses of a stone wall. In their hands was the Valley Quest guide to Questing in the Upper Valley region of Vermont and New Hampshire. Playing dumb, I asked them what they were up to.

"Oh, we're on a treasure hunt," they laughed. "Haven't you ever seen this book? It's one of the best things we've ever come across. We moved to the area about a year and a half ago and we use the Quest guide to help us figure out what we're going to do each weekend. The places we go to are great for our two-year-old and four-year-old and the Quests keep us adults challenged. Moreover, we would never find our way to all these wonderful, off the beaten track locations without our Questing guide. It's given us a whole new sense of place."

You too can be like Godfrey. As searchers follow your Quests, they'll quietly thank the genteel and smart chef who provided the sumptuous—and free—repast that you've laid out before them. Bon appetit!

Questing

Introduction: Place-Based Education

PLACE-BASED EDUCATION is learning about, learning from, and learning within the context of where we are. Our places, whether they are in the city or the country, are filled with grasses, shrubs, trees, amphibians, insects, birds, mammals, reptiles, and other humans. Our places are filled with cultural elements, too: roads, buildings, bridges, orchards, canals, dams, parks, museums, reservoirs, and wastewater treatment plants. All of these elements nest together in a single place we call home. Each one, no matter how small—from the rusty railway spike to the track of a doe—tells a tale of inhabitation. We can follow the railroad bed to the ruins of the old roundhouse or backtrack to the place where the doe browsed on new hemlock growth and then slept the night before. We can study geology where the highway slices through bedrock or learn more about biology and chemistry through our exploration along ponds, rivers, and streams.

Right now, in our community, there are outdoor classrooms offering free courses in agriculture, anthropology, biology, botany, chemistry, geology, history, sociology, and structural engineering. The place we live need not simply serve as the background setting or stage for our lives; rather, this place—*our* place—can serve as a primary resource for education, entertainment, nourishment, inspiration, and lifelong learning. No matter where we live, the natural and cultural landscapes of our communities offer an array of teachings that, taken together, tell the particular tale of how life has unfolded *here*.

Place-based education is about discovering the stories of home and using them as the foundation for learning, community building, and stewardship. Its core practices include

- ➤ investigating our natural and cultural surroundings;
- ➤ adopting specific settings or issues in the community for in-depth study, as well as studying regional, national, or global issues through the lens of the local;
- ➤ deepening understanding through the study of local primary and secondary resource materials;

➤ meeting, sharing perspectives with, and learning from our neighbors, including elders, wisdom-holders, and partnering community organizations;

➤ sharing these "lessons learned" through community service.

Wandering through our landscapes is a rewarding activity. We can discover or remember hidden or forgotten community assets. We can find overlooks as well as tiny nooks, places where *what we see* and *how we see* changes and we gain a fresh perspective.

Without embracing what is out there, we suffer a great loss. Without a sense of connection, a familiarity and intimacy with our places, we have no sense of the implications of our lives or actions. As adults, our motivation to participate in community affairs, and to act on behalf of our communities and the natural environment, comes directly from a strong sense of place. As Vermont author John Elder noted in *Reading the Mountains of Home*, "Love is where attentiveness to nature starts, and responsibility towards one's home landscape is where it leads."

The importance of connecting to places is as clear for children as it is for adults. Gregory Smith, professor of education at Lewis and Clark College in Portland, Oregon, writes:

Over the past decade, educators from New England to Alaska have been relocating the curriculum away from generic texts to the particularities of their own communities and regions. This process has been accompanied by the adoption of instructional practices drawing heavily on student initiative and responsibility—as well as the talent and expertise of adults outside the school. The results have included higher levels of student engagement, greater commitment to public education, energized and excited teachers and principals, and a renewed sense of what there is to value in the local.[1]

This book is about a place-based education model called "Questing." Questing is creating and exchanging treasure hunts in order to collect and share your community's distinct natural and cultural heritage, your special places and stories. Each Quest is a specific treasure hunt focused on a particular community story, environment, or character: here's where the first trading post was; there's where you can see spotted salamanders in the springtime; on those buildings—if you look carefully—you'll see water stains from the great flood. A Quest's clues and map lead to a hidden treasure box, but the clues are treasures, too. The clues are the signs—the evidence, really—that one can learn to "read" as the pieces fit together to tell the story of a community. A Quest, through rhyming, riddle-like clues and a hand-drawn map, becomes a playful game in which what was heretofore hidden begins to become clear. Questing is local and organic; it is authentic and interdisciplinary; it is personal, collaborative, and intergenerational.

Questing: A Guide to Creating Community Treasure Hunts is intended for a broad audience of individuals, families, organizations, school groups, youth groups, civic groups, museums, and nature centers, among many others. Anyone can make a Quest. And the Quest itself is a flexible format, adaptable to a wide variety of contents and purposes. No matter where you live, we believe that Questing can help you learn to see more clearly, share, and enjoy the riches that are already there.

Chapters 1 and 2 introduce the core concepts and background history of Questing. The next four chapters walk individuals, organizations, and classroom groups through the preliminary phases of a Quest project. Chapters 7 and 8 focus on research—first out in the field, then indoors utilizing community source materials. Chapters 9, 10, and 11 move the reader through the creative process of writing clues, making maps, and preparing the treasure box. The final chapters help you transition from a single Quest to an ongoing community Quest program.

Wherever you live and wherever you look, there's treasure out there. Explore, enjoy, and happy trails!

1) The Joy of Treasure Hunts

AH, THE LURE OF TREASURE HUNTING! Somewhere out there is a hidden surprise. If you apply your best thinking and most clever strategies, you will find it. The hunt was created to entertain you—to lure you into a mystery, a story, an exploration of some hidden, magical world. All you have to do is solve the clues and decipher the map, and the treasure will be yours.

Treasure hunts have served all kinds of purposes, across the generations and around the globe. Before we examine the ways that Questing uses treasure hunts to accomplish community-building, educational, and recreational objectives, let's glance through treasure hunting's rich history.

OPENING EYES

For thousands of years, our ancestors moved across the landscape in small groups, reading and remembering its signs and stories, whether they were looking for grazing fields, wild edibles, water holes, or signs of danger. These skills seem to be hardwired into our brains and our physiology. Our upright, biped posture, with head held high, allows for sweeping vision. We are sensitive to contrasts in color, light, texture, and motion. We carry the memories of walking with eyes wide open in our bodies and minds.

Whether you are searching out the license plates of all fifty states on a road trip or counting hawks roosting on fence posts and utility poles, there is something natural and joyful about the attentive gaze. And this gaze is heightened by our experience of motion. While we move or travel, we look out at what lies ahead, gazing in anticipation at the coming unknown.

When we begin to look for something particular, our perceptive abilities become focused. We feel more alive as we look around; we seem to notice more than usual. The things we notice—the outer elements (that tree, that rock), the inner bodily feelings (changes in temperature, a subtle breeze), and the feeling of a relationship between the two—work together to make us feel more *present* in the place and moment. Our focus is pulled out from the drifting undercurrent of thoughts and ideas and into the living, breathing world of direct experience.

1.1 Esther Currier Quest, Elkins, New Hampshire. Photograph by James E. Sheridan. Treasure hunts engage our senses, and treasure hunters are teased forward clue by clue. Here, clues lead down to an inactive beaver lodge at the edge of a pond.

Hot on the trail—a few clues into a treasure hunt—there is the excitement of discovery. Each clue is one more step toward the completion of a life-size puzzle. The further we move into a hunt, too, the greater our anticipation for the next clue becomes. Sometimes the heartbeat, quite literally, races toward the treasure.

GIVING GIFTS

There is something wonderful about making treasure hunts, too. They are a gift we can give to the world. Annie Dillard writes in *Pilgrim at Tinker Creek:*

> When I was six or seven years old, growing up in Pittsburgh, I used to take a precious penny of my own and hide it for someone else to find. It was a curious compulsion; sadly, I've never been seized by it since. For some reason I always "hid" the penny along the same stretch of sidewalk up the street. I would cradle it at the roots of a sycamore, say, or in a hole left by a chipped-off piece of sidewalk. Then I would take a piece of chalk, and, starting at either end of the block, draw huge arrows leading up to the penny from both

directions. After I learned to write I labeled the arrows: SURPRISE AHEAD or MONEY THIS WAY. I was greatly excited, during all of this arrow-drawing, at the thought of the first lucky passer-by who would receive in this way, regardless of merit, a free gift from the universe.[1]

TREASURE HUNTING WITH FRIENDS AND FAMILY

In Poland, *podchody* (literally, "to go under," or to sneak up on) is a favorite pastime of adolescents and teenagers. Polish emigrants have carried this tradition around the world, binding communities across place and generations. In the Adelaide hills of Australia, *podchody* often serves as the centerpiece of an evening's entertainment for youth from the same Polish church community. As Daniel Jantos describes it,

> We usually did it at my parents' place. We had eight and a half acres—mostly apple orchard, some cherry, some open. We'd generally have about twenty to thirty kids, mostly teenagers. The older people loved to see us do *podchody* and they would make tons of food for us—Mum would always make a cake if she knew we young people were getting together. People would park by the house and hike up into the field, about three or four hundred feet up. After we'd eaten, we'd divide into two teams and one of the teams would head off, leaving clues for the other group to follow.
>
> After about half an hour, the other group followed, looking for arrows made with sticks or drawn in the dirt. If you found a crossed arrow sign, that meant that a clue was hidden somewhere nearby. Clues not only told the team where to go next but provided a lot of the best fun of the evening. They were full of in-jokes referring to the other people in the game—the wittier the better and sometimes a little on the rude side! We usually tried to make them rhyme, too. The team wrote the clues as they walked, so the notes got quicker and quicker as the second group began to catch up. Sometimes the first team would double back on their trail and hide to spy on the second team as they read a clue, just to see if they got the joke!
>
> We always kept an eye out for interesting places to lead the group the next time we played, like a big old shed in our neighbor's orchard. We knew all the farmers well, and if we told them ahead of time, they'd never mind. We'd sometimes lose people along the way, you know, couples might drift off, but in the end it would all lead to a big fire ring where the other group would be waiting with a hidden party—more food and a bonfire ready to light. People would go get their mandolins and guitars and we'd all sing and tell jokes late into the night. We had some wonderful times doing that![2]

Exploring the landscape with friends, trying to trick each other, lingering out from daylight through dusk and into the night—it is precisely these kinds of interactions that foster and nourish a sense of place, community, belonging, and wellbeing.

Treasure hunting has been taken to some of its greatest heights by loving, creative parents for their children's birthday parties. Linny Levin, a noted environmental educator and former coordinator of the Valley Quest program, coaxed her six-year-old son, Casey, and his friends up a steep two-mile trail using a clever, homemade treasure hunt. She hid handmade wooden and leather elves in rock crevices on the trail up the mountain. The children's eyes became so accustomed to noticing small telltale clues that they had no trouble locating the rock cairn on the summit, where party favors of leather treasure pouches awaited them.

And what better way to entertain guests than with a treasure hunt? Here's an excerpt from one created by Susan Clark. It charmed guests at a wedding shower with a tour of house and garden and clues hinting at lively stories from the bride and groom's past.

> They have them in April
> They have them for babies
> They have them when people take vows.
> Mostly they're indoors
> (but this one is out)
> In fact we're having one now!

Treasure hunts don't require a large audience, either. They work just fine one on one. Delia was once delighted by a Valentine's Day treasure hunt created just for her by a college sweetheart. Starting with a cryptic card in her campus mailbox, a succession of hidden Valentine cards led all over campus, one taped onto the underside of "her" table at the library, another tacked near a rendezvous spot in the student union, and so on. The clues led to a strawberry sundae at the dairy bar and then back to a beautiful homemade Valentine at a desk in a shared campus office. Prospective Casanovas take note: treasure hunts can be effective!

TREASURE HUNTING AS CIVIC ENGAGEMENT

Nonprofit and community organizations have hit on many ingenious ways to leverage the power of treasure hunting toward accomplishing their civic-minded goals. *Tracking the Dragon* is one example, calling itself "A Puzzle with a Pur-

1.2 Hunt for the Elves. Photograph by Simon Brooks. Homemade wooden and leather elves were tucked in rock piles and stone walls to mark the route of a birthday trek up a mountainside.

pose." This environmental game was invented by Wild Olympic Salmon, a community-based nonprofit organization in the Pacific Northwest dedicated to healthy watersheds for wild salmon. Participants are lured into "celebrating the hydrologic circle" by exploring interesting spots throughout the watersheds of Jefferson County, Washington, while they search for twelve well-concealed bronze dragon footprints. These treasure hunts, created by teams of artists, poets, and sculptors, were originally published in local newspapers, one each month, building a froth of anticipation in eager trackers. As local resident Greg Benton reported, "We've gone without food, even TV, to go Tracking the Dragon!"

The clues are now gathered in a book, complete with intricately bordered blank pages where treasure hunters can make rubbings of the tracks they've found. Olympic-area resident Dennis Kelley's comments on his dragon tracking experience could aptly apply to many successful community educational treasure hunts: "What I have enjoyed most about *Tracking the Dragon* is the opportunity to become, not childish, but childlike, once again. For a brief period of time, the distinction between adults and children has blurred, and we're all, the eight of us, childlike friends pursuing a year-long gigantic scavenger

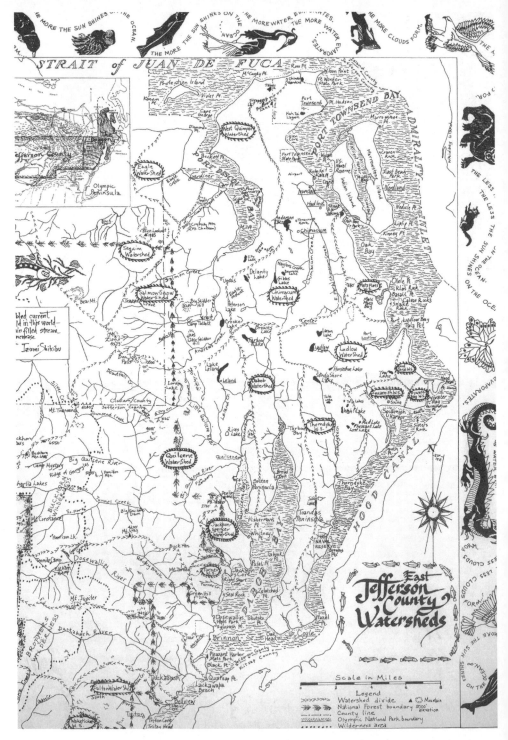

1.3 *Tracking the Dragon* is an environmental game and watershed tour conceived of by Wild Olympic Salmon. Hunters search out twelve bronze "dragon tracks" hidden in Jefferson County, Washington.

Tracking the Dragon is an original concept created by Sara Mall Johani and organized under the auspices of Wild Olympic Salmon, a non-profit community organization in Jefferson County, Washington. *Tracking the Dragon* and its replication *Tracking the Thunder-bird* in neighboring Kitsap County was funded by Puget Sound Water Quality Authority (now called Puget Sound Water Quality Action Team). Thirty-six community artists, writers, and sculptors were engaged to create the guide for *Tracking the Dragon*.

hunt. We've been able to rekindle our sense of wonder, spend some good times with our families and friends exploring this wonderland in which we live, and we've all had a tapestry of rich experiences in the process."[3]

In New Hampshire, local environmental educators worked with Antioch New England Graduate School professor David Sobel to build public interest in the natural and human history of an underutilized state park through *Pisgah Treasures.* Clues and riddles, released over time in local newspapers, required both research in town libraries and extensive tromping to all corners of the 13,000-acre park. The first year, local treasure hunters solved three months' worth of clue writing in less than twenty-four hours. The next year, the group came back with a more challenging set of clues that eventually took local treasure seekers four months to solve.

Helen Peterson, part of a two-couple team that finally broke the code, said, "It started out as a lark and sort of became an obsession. We were going to find it and that's all there was to it." She and her husband and friends braved sun, bugs, and rain every weekend from June through October to find the eleven clues that pieced together the words "School Eight," which was the site of an old school near which the treasure was hidden. Along the way, the educators who organized the hunt accomplished one of their side goals as well: they raised the visibility of a new high school social studies curriculum they had just written, called *Yankee Lands,* which held the answers to several of the clues.

TREASURE HUNTS FROM THE REALM OF COMMERCE

Entrepreneurs have found myriad ways to translate the allure of treasure hunting into commercial profit. Visitors to Disney World may be familiar with the *Adventureland Challenge* that sends vacationers in search of the "Idol" hidden among the props and shrubbery of the Magic Kingdom. While we are assured that the Idol will not be hidden inside shops, attractions, waiting areas, or restaurants, we don't have the same guarantee for the clues:

> Hold the merchandise to your ear,
> you'll hear the sounding sea.
> Get your next clue from the clerk here
> and on your way you'll be.

We're sure most parents aspire to taking nothing but clues and leaving nothing but footprints while inside the string of shops, but our guess is that

Disney is hoping otherwise! When all the clues are gathered and assembled, puzzle fashion, their backsides reveal the Idol's secret hiding spot.

Many books and computer games utilize the tried and true treasure hunt formula as well—for example, *Where's Waldo, I Spy,* and *Blues Clues.* The creators of these games have found ways to successfully get inside the psyches of their users with the seduction of alternative worlds, as evidenced by twelve-year-old Toben Traver's explanation to us, one day, of the computer game *Quest for Glory.*

> There's a hero and you name him and you control him—it's sort of like you *are* him. He's trying to save the city of Silmaria on the Island of Marete from some big problems, like an army of evil soldiers, monsters, and a dragon. There are tons of secret clues and if you solve them, you can save the city. Some are hidden inside doors or chests, so you either have to figure that out or else check in every single one.

While these commercial ventures may look like serious competition for the attention of the youngsters and adults you're hoping to lure into making a Quest with you, another way to look at it is that members of your group may know more about treasure hunting than you expect. The challenge is to leverage this expertise into accomplishment as a Questmaker and to make that Questing so much fun that traipsing around the neighborhood becomes serious competition for computer games.

ENTER . . . QUESTING!

So how does Questing fit into this world of treasure hunting? While certainly a heap of fun, Questing was developed to accomplish an important mission: strengthening the vitality of local communities. By encouraging young and old to actively research, explore, and have fun in their local natural and built landscapes, Questing fosters a sense of place, sharing the distinctive stories of local heritage and environment. At the same time, it helps preserve the natural and cultural heritage and strengthens social capital, that invisible web of trust and reciprocity that holds a community together and keeps it healthy.

Quests are inherently local and inherently noncommercial. They can be created by school classes, families, conservation commissions, arts councils, scout troops, chambers of commerce, science museums, nature centers, historical societies, land trusts, local government, state and national parks, informal groups of friends, and anyone else who wants to celebrate the place they call home. Quests have proved popular in rural and urban areas alike, focusing on everything from the escapades of local heroines to a closeup look at carnivorous plants in a local bog. Questing is contagious and each new program has taken on a life of its own, breeding all kinds of local variations and strange traditions.

2) The Story of Questing

QUESTING EMERGED FROM OUR SUSPICION that a great way to build people's sense of place would be to invite them out into the landscape to play. You know, *play:* bend down and smell moss, explore a new neighborhood or go deeper into the forest, look behind and under and up at the tippy-top of things, solve riddles, and learn the juicy details of local lore. We had faith that if we could make the game fun enough, we could lure people away from their home entertainment systems and out into their communities. And we believed that if we could do this, it would be a good way to inspire people to become active stewards of their community's natural and cultural environment. As John Elder points out, "adventure is the best starting-point for education and citizenship alike."[1]

OVERVIEW OF QUESTING

Each Quest is a permanently installed treasure hunt that utilizes poetic clues, hand-drawn maps, and occasional sketched hints to guide self-appointed Questers gently and playfully through an environment. Quest clues are usually written in lively rhyming verse and often tell a story about a local place, character, or phenomenon. Quests end at a hidden box, where one can sign in and collect the impression of a custom-designed rubber stamp that depicts an image related to the site. The treasure box often reinforces the story theme, containing small caches of art supplies, science equipment, or printed materials that help visitors to better understand precisely where they are. Questing enthusiasts usually carry their own handmade stamp so they can personalize their sign-in book entries in each box they find, adding a bit of spontaneous poetry if they are so inspired.

Each Quest is built up from three essential elements: a set of clues, a map, and a treasure box. This sample clue, excerpted from the Trustom Pond Quest, demonstrates how clues can be used both for giving directions about movement and for telling a story.

2.1 The Beaver Meadows Quest. Photograph by Jon Gilbert Fox. The Beaver Meadows Quest guides visitors through a historic settlement, pointing out the one-room school-house, cider mill, blacksmith shop, church, parsonage, and post office. Some of the buildings are gone, but the stories live on through the Quest.

> Go straight ahead as you come out in the sun
> Walk a few minutes and then for some fun
> Look on your left for a windmill there
> How many legs hold it up in the air?
> When this land was a farm it was used to pump
> Water for fields and animals plump
> And one time sheep were raised on this land
> Now they are gone and nature has planned
> To replace the fields with bushes and trees
> And nature can do whatever it pleases.

Once created, Quests may be gathered into a booklet that is updated regularly or printed as a series of individual flyers sharing a similar graphic look. They may also be featured as a regular column in the local newspaper. In more established programs, Quests can be printed into a bound book and sold in bookstores, inns, and recreational equipment shops, with revenues supporting

As you go on the Quest, I hope you will see
Forest disturbances, there are more than just three.
Clues of blowdown, logging or fire you may find
Because our forests are changing all the time.

Walk north up the road after parking your car.
Turn on the trail that is left and not very far.

Leave ruins of pond and foundation of mill,
Then cross the bridge and walk up the hill.
Pick up needles. They'll be on the ground.
Needles together in two's and five's will be found.
The two's are red pine, with rough scaley bark.
Were these trees planted in rows in woods so dark?

A red oak up ahead has a story to tell.
The four trunks grew after its ancestor was felled.
So pass #6. Look left after 100 feet more,
then imagine the diameter of this tree's original core.
> To calculate: Note the center of each trunk.
> Connect them in an imaginary circle near the ground.

Cross the stream at #7 and on the right you will find
a "pit and mound" that will boggle your mind.
The mound was made from decomposing roots and debris
and the pit shows where a tree used to be.
So if you find a pit right next to a mound,
then evidence of a very old blowdown you have found.
> The tree fell to the southwest so it was a
> northeasterly wind that blew the tree down.

On the left after 15, a poplar was scarred long ago.
Was the basal scar from fire or log skidding? Can we know?
Note that a collection of leaves on an uphill side
provides the fuel for a fire to thrive.

Stop to look at the rock walls in the canyon nearby.
Continue left on the trail and cross where a bridge of stone lies.
Bearing from a multi-trunked birch near a pit and mound,
go 230 degrees to a multi-trunked tree.
Here, the box will be found.

2.2 Grand Canyon of Norwich Quest map. Maps work together with the clues to lead visitors to the hidden treasure box. Maps can mark the starting place of a Quest, describe more generally the terrain explored, or help clarify the treasure box location. *Courtesy of Valley Quest: 89 Treasure Hunts in the Upper Valley.*

2.3 Woodstock Village Green Quest C treasure box. Photograph by Jon Gilbert Fox. Treasure boxes can be a sandwich container or a utility box, or perhaps not a box at all but instead a water bottle or a coffee can. Treasure boxes are hidden indoors and out. They often lurk in public spaces, right under your nose, disguised as everyday objects.

improvements to the Questing program. A fully developed Quest program might include a group of Quests, a cycle of annual events, a newsletter or website, and a community of volunteers that support the program.

THE HISTORY OF QUESTING

Questing arose out of a 150-year-old tradition called Letterboxing, popular in the Dartmoor region of England. Letterboxing began in 1854, when a Victorian gentleman named James Perrot of Chagford walked the wet and boggy nine miles out to Cranmere Pool and stuck a bottle containing his calling card in a poolside peat bank to claim credit for his accomplishment. Subsequent walkers began leaving their cards in the bottle as well, and by 1908, when the bottle had been replaced by a tin box, the visitor's book it contained had received 1,741 signatures. This box allowed for the possibility of leaving something larger than a business card, and visitors began leaving postcards, stamped and addressed to themselves, for subsequent visitors to post, for the fun of seeing when and from where their postcard would arrive. Hence the name "Letterboxing."

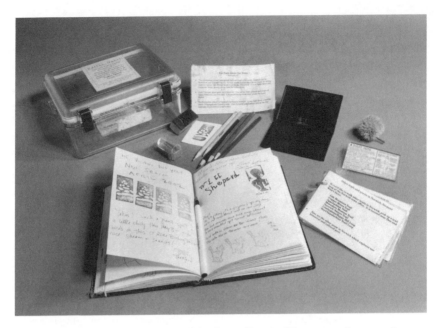

2.4 Treasure box contents. Photograph by Simon Brooks. The treasure box contains a sign-in book, a pencil or pen, a custom-designed rubber stamp and stamp pad, a short essay about the site, and additional materials that help visitors learn more about where they are.

Another box was placed on the moor at Duck's Pool in 1938, and over the years the number of boxes continued to grow steadily. In 1976 a local walker produced a souvenir guide map to the fifteen Letterboxes then in existence. That kicked off a rash of new box installations by both clubs and individuals. As the tradition shifted away from the leaving of calling cards, people began looking for something more personal than just signing their names in the sign-in books. They took to carving handmade stamps out of rubber or linoleum that depicted some part of their personality or family history (sometimes quite obscure!). Personal stamps are now such a firm part of the Letterboxing tradition that a stamp-making cottage industry has become established in the region. Many Letterboxing enthusiasts still design and carve their own individual stamps. Most have also become quite proficient at creating verse to accompany their stamps in the sign-in books. The verse ranges from tender to bawdy, the former exemplified by this verse, presumably written with apologies to Wordsworth:

2.5 Stamp impressions collected from Dartmoor's Letterboxes. There are now more than 10,000 treasure boxes hidden in and around Dartmoor. Visitors search the moor to collect the coveted impression of Benchtor or the Peat Cutter of Dartymoor. *Courtesy of 101 Dartmoor Letterboxes, by John Hayward and Anne Swinscow.*

Daffoboxes

I fluttered aimless as a kite
That's come to grief among the rocks
When lo! I spotted something white
That looked just like an ice cream box,
Beneath a boulder, pushed well back
And stuffed inside a plastic sack.

So crouching on the muddy ground
I pulled it out to have a look;
Released the lid, and inside found
A stamp, a pad, a soggy book.
As these I saw with just a glance
My heart with joy began to dance.

Now oft when o'er the Moor I pass
In vacant or in pensive mood
These boxes tumble on the grass
To break a spell of solitude.
And then my heart with passion rocks
And dances with a Letterbox.[2]

As the number of boxes grew, some of those new to the game demonstrated a lack of understanding of the importance of protecting the antiquities and fragile natural areas on the moor. Dartmoor National Park officials cracked down and attempted to institute a policy prohibiting boxes on the moor, except for the two original boxes at Cranmere and Duck's Pool. A hot debate ensued, attracting headlines in the late 1970s like "War on Little Boxes Littering Dartmoor." Happily, the resolution proved to be an agreement by all Letterboxers to abide by a code of conduct prohibiting disturbance of the natural landscape or any of the cultural resources.

This allowed for the explosive growth of this odd but engaging pastime. Today in the Dartmoor region everyone from toddlers and teens to parents and pensioners dons their Wellies on a regular basis and, using maps, compasses, and esoteric clues, traipse the moors in search of hidden boxes. The current *Catalogue of Dartmoor Letterboxes* lists around 4,300 boxes known to be presently out on the moor, with each box owned and maintained by an individual or a club. The catalogue also includes listings of boxes in the public buildings and pubs surrounding the park, as well as "traveling boxes" carried by

Letterboxers as they wander. When these are added to the boxes whose time has expired in the catalogue but which may well still be in place on the moor, the number climbs to over 10,000. Anyone who has visited and collected stamps from one hundred bonafide Letterboxes is eligible to join the "100 Club," which meets twice yearly near Dartmoor on the clock change days. The club's membership currently includes over 13,000 hearty walkers.

QUESTING COMES TO THE U.S.

David Sobel became fascinated by Letterboxing while on his sabbatical in England in 1987. He reports that he kept seeing people tramping about "with the intensity of Eagle Scouts," studying small tattered sheaves of paper. After repeated inquiries he was finally able to find someone who was willing to clue him in, and he was off Letterboxing.

When Delia was looking for a concrete way to engage youth in their communities through Vital Communities, a regional nonprofit organization she had co-founded in the Connecticut River valley of Vermont and New Hampshire, David described what he had seen to her. He added that he had been looking for an opportunity to experiment with treasure hunting as an educational program. With his help, Vital Communities and Antioch New England Institute built on the English tradition on this side of the Atlantic by developing the Valley Quest program. Later, Steve was hired as the Valley Quest program coordinator.

Together we transformed the Letterboxing tradition from a focus on recreation into an integrated experiential curriculum exploring sense of place. We field-tested the curriculum with public and private schools, home schools, nonformal educational organizations, youth groups, and an array of community partners. In 1997 the Valley Quest program received the Montshire Museum of Science's Walter Paine Award for Excellence in Science Education. After several years, the growing body of Quests proved to be a great tool for building community and sense of place across the region, in addition to its original educational value. In the Upper Valley region of Vermont and New Hampshire, where Valley Quest has now been underway for eight years, the most recent edition of the Valley Quest map book includes eighty-nine Quests. More than fifty teachers, 750 students, and 150 community members contributed to the creation of these Quests, which stretch across thirty-one towns. Another seventy-five Quests have been collected in anticipation of publishing a second volume of Quests in summer 2004.

Increasingly, people leave their trips to the Upper Valley with more than just maple syrup and photographs. The Questing seed has been planted, and partners like the National Park Service Rivers and Trails Program and the Appalachian Mountain Club have stepped in to support the growth of Questing programs in other regions.

QUESTING: AN EDUCATIONAL ADVENTURE

While Quests are often created informally by inspired individuals, they can also serve as meaningful project opportunities in which students and organizations contribute to the vitality of their communities. In this case, work might begin with a study of community maps, both historical and modern, and an assessment of local places of special interest, beauty, or significance. Once a group has chosen a site and a theme, members conduct field research and interviews with community members to learn more about their subject. They make maps and write poetic clues to pass on some of the most important and enduring things they have learned. Finally, they design a stamp that embodies their story, and they hide their treasure box, packing it with a list of fun facts about the site or suggestions and materials for activities to do there.

Teachers who incorporate Questing into their work find that it naturally integrates the curriculum. The borders between field science, mathematics, social studies, language arts, and art blur as students write clues based on observation, collect data in the field, sketch from nature, and interview local experts. Everyone from college professors and high school honors faculty to special education teachers and elementary educators has used Quests successfully as thematic units in the curriculum. Questing is also ideal as a service learning project. The students' satisfaction and pride in creating a product of value to the greater community lays a solid foundation for their future engagement in civic life.

Community members appreciate the effort put into a Questing program, as this comment from a participant in Boston's South Shore Quest program demonstrates.

> I thought the Quests would be a fun way to explore the outdoors on some of these summer days. But yesterday we found that it is also a good way for six- and nine-year-olds to work on everything from directions (left, right, north, south) to map-reading to cooperative problem solving. (There was a certain amount of competition that had them racing off to find the clue before I'd finished reading it.) I see now it is also an opportunity to connect to our com-

2.6 The Blacksmith Bridge Quest makers. Photograph by Steve Glazer. Cornish, New Hampshire, students adopted, studied, and mapped the Blacksmith Covered Bridge in the process of building skills in geography, history, and language arts.

munity (and neighbors!) in a way I never imagined. Thank you for putting your time and creativity into this effort. We're looking forward to completing the rest of the Quests over the summer.

NATIONAL INTEREST IN BOX-HUNTING

One evening, Delia was participating in a salon hosted by SWEEP, Vermont's Statewide Environmental Education Programs alliance. The event featured great food and a rich conversation about place-based education. Delia was waxing prosaic about her latest enthusiasm, Questing, and its roots in Letterboxing. As the meeting drew to a close, Chris Granstrom, a writer for *Smithsonian* magazine, approached her, wondering where he could find out more about it.

Chris's article, which appeared in the April 1997 issue of *Smithsonian,* gave birth to what is now a national Letterboxing movement in the United States. North American Letterboxers are organizing and communicating nationally (via the Internet) and are quickly approaching their goal of hiding Letterboxes in every state. Like Questing, Letterboxing engages a loose affiliation of people who are committed to keeping their organization and their web presence nonprofit and commercial free, with open access to Letterboxing clues. There is no membership required—private contributions cover the costs of operating the site. While Questing tends to involve a stronger mission focus on place-based education and community building, and a tighter local network, Letterboxing tends to involve a stronger recreational focus and a loose national network. Both share the goal of getting people outdoors and exploring. There is, increasingly, a crosslisting of Questing and Letterboxing sites, taking advantage of the synergy between the programs.

Another treasure hunting program gaining in popularity is *Geocache*. This high-tech game is similar to Questing and Letterboxing in its focus on clues leading to treasures stashed in unlikely places. The difference here is that the caches are all identified by coordinates and the treasures are found using electronic handheld GPS devices. Also, instead of finding a box with a logbook and a stamp at the end of your Geocache, you'll find a collection of objects left by the people before you (books, CDs, figurines). You are allowed to take something, but only if you leave something in return.

USE QUESTING TO ACCOMPLISH YOUR GOALS

Good Quests offer the opportunity to extend the learning of the Questers and encourage them to explore the community in more depth as part of their Questing experience. These excerpts show what others have encouraged the folks on their Quests to do:

Have a picnic (from the Jonathan Wyman Sawmill Quest):

> Feel peace surround you as you walk in,
> For you are someplace you have never been.
> A brook and a waterfall are things you will hear,
> A birdsong, leaves' crackle, a glimpse of a deer.
> A nice place to picnic would be right here.

Check out a nearby view (from the Balch Hill Quest):

> After finding the treasure, continue on to the top.
> There's a beautiful view that will cause you to stop.
> Nearby, too, is an amazing oak,
> Seventeen feet around—this isn't a joke.
> See Mt. Ascutney from the top, Baker Tower too.
> And on your way down, follow blazes of blue.

Come again in different seasons (from the Lonesome Pine Quest):

> Ah . . . there is so much more to this story,
> Like the hillside in fall in all its glory.
> So many mysteries waiting to be solved,
> In a place so far away from the maddening crowd.
> Let the peace feel its way in . . . 'cause there's so little that's loud.
> Spring, summer, fall—which time is the best?
> Come again and again, and experience the rest.

Look for wildlife (from the Enfield Rail Trail Quest, near a historic Shaker site):

> To the right of this bank is a rich habitat,
> A cool wetland home for ducks, in fact.
> Explore a little and you may see
> Frogs, turtles, or herons . . . lucky thee!

Read the interpretive signs (from the George, Frederick, and Laurance Quest):

> Your next clues are inside this Visitor Center,
> Follow the signs and then you should enter.
> Go to the sign that's straight ahead with great speed,
> There you will find some letters you need.
> Stick with this sign and read it through
> You'll learn a lot and have fun, too.

Eat at a local restaurant and support a local business (from the Bellow's Falls History Quest):

> If you're looking for a place to plop yourself down,
> Walk north to the clock tower in the middle of town.
> This cafe's espresso is surely the best
> So that is the place we want you to go next.
> The sign on the side looks just like a moon,
> So step right inside and buy coffee soon!

Get your feet wet (from the Spears Dam letterbox):

> A grassy bridge you will shortly find.
> Though not required, if you don't mind,
> Go in the stream, in awe you'll stare:
> Some pretty incredible stonework there.

Reflect in silence (from the Polly's Rock Quest):

> Nestled among the trees and quiet, feel a release!
> Find yourself marveling at the solitude and peace.
> Polly, too, found these woods her special place,
> So special she gave them to be saved, with grace.

Do more Quests (from the North Main Street Quest):

> Read the stories, stamp your book,
> Walk around and take a look.
> Hope you enjoyed 4-Parker's Quest.
> There are more! Do the rest!

Whatever your other goals, we're sure your community will appreciate your efforts and agree with British Letterboxing enthusiast John Hayward's sentiments:

> We thank the many people
> Who hide these boxes small.
> We spend our lives out hunting,
> Yet cannot find them all,
> For some are hid in clitter
> And some in peaty banks,
> But whereso'er we find them
> Accept our grateful thanks.[3]

3) The Spirit of a Place

THE VERY BEST QUESTS capture the spirit of a place. Sharing this spirit, however, requires that we've had this feeling or discovery experience ourselves. If we hope to find the spirit of a place, we need to learn how to see the details—and discover hidden stories.

SEEING DETAILS

Steve lived for a while in southeast Arizona, in a town called Patagonia. New to the Sonoran desert—and the whole of the southwest—he would head out in a new direction each morning, exploring. Walking in the early morning light was invigorating and offered up wonderful things: the spiraling architecture of a cholla skeleton; the hidden crystals of a geode; the surprising red flowering of ocatillo after the rain; marching male tarantulas out searching for mates; a green rock outcropping that watched over the dry river valley like a giant frog.

Walking in the desert, one learns that it is often easiest to travel the washes, where the water has carved a path for you. Meandering through river-run sand and over rocks is easier then competing with catclaw, crucifixion thorn, and mesquite.

One day Steve was out in the wash with some friends, Todd and Blake, at a place he had passed through dozens of times, where his daughter Kayla had spent a lot of time collecting rocks. Now everyone knows how to look at the ground and collect rocks, but Todd knew how to see and recognize something more. He reached down for a single rock that caught his attention: "Flake!" he cried.

"What's a flake?" Steve asked. Rather than answering, Todd quickly scanned the ground, picked out two stones, and whacked one against another. In just a few swift strokes, Todd had turned a stone into a primitive cutting tool. "A flake," he explained, "is the byproduct of flintknapping and toolmaking. It's like the sawdust, or the leftovers. This other piece—the one you are chipping away at—that is the core; and eventually it becomes a tool. Someone, once upon a time, was making tools nearby. These are rhyolite flakes."

3.1 Frog Rock, Harshaw Creek, Arizona. Photograph by Bob Roberts. Peeking over oak and mesquite toward Patagonia and Mt. Wrightson, Frog Rock has been sitting here for a long time and has witnessed many changes in the landscape.

Todd walked a few steps further. "Another one," he called. And soon, the whole group could see them—they were easy to see, easy to find. As they looked around carefully, they decided that the source of the flakes seemed to be a tiny eroded gully cutting down through a small, rounded hillside into the wash of Harshaw Creek.

As they headed out of the sand and up into the gully—through tall, blonde sacatone grass and the green-gray mesquite grove—the density of the flakes began to increase. At least one flake was noted with every single step of the ascent. They were beginning to find rudimentary tools as well: blades, hammers, "cores" that had split the wrong way and been abandoned. Forty feet or so up the ravine, the land flattened out. "Here it is. Awesome. A chipping ground," Todd said.

"Shard!" Blake cried out. She lifted up a small, clay-colored object the size of a thick, broken potato chip, smooth on both sides and slightly crystalline along its edge. "Pottery!"

Blake worked as a cook on Grand Canyon rafting trips and was used to seeing ancient artifacts. From where the group was now standing, everyone, *anyone*, could bend at the waist, peer toward the ground, and learn to see pottery

3.2 Scraping tool and pottery shard from Harshaw Creek, Arizona. Photograph by Simon Brooks. Once you learn to see the multiple jagged faces of a core or tool, you will observe them everywhere.

shards. Until someone points out these cultural relics, however, they are easy to miss. Once one learns how to see, however, signs of the past are everywhere.

Anywhere the surface was disturbed, there were artifacts: in depressions, on the sides of small mounds, among the roots of the dusty mesquite trees. The group found projectile points, pottery shards, jewelry, and mano and metate (mortar and pestle grinding tools). Most were left on site; a few representational samples were taken for study and interpretation.

Some shards were plain. Others were glazed. Still others were decorated monochromatically. A few were even polychromatic: pots painted with a base color of white and then decorated with designs. One magnificent piece featured a geometric design in red and black over a creamy white background. Later, Alan, an archeologist, came to visit the site. He estimated that it held as many as 250 artifacts per square meter. This place had transformed from being a nice place to collect rocks into something else entirely. He also pointed out that the patches of sacatone grass were likely to be sprouting out of disturbed ground—perhaps the sites of old pit houses or other dwellings.

Based on the density of the pottery and the diversity of the pottery styles,

Alan felt sure that this site was at least a medium-sized settlement, and that it was inhabited periodically over the course of several hundred years, perhaps from A.D. 800 to 1400. The changes in color and design corresponded to the changing dates of inhabitation.

So Todd taught the group to see flakes. Blake taught them to find shards. And Alan taught them to envision a settlement.

When they looked around and saw the land itself, they thought, "Why would someone choose to settle here? Why make this place a village?" They then climbed up adjacent Moon Hill and looked down on the setting. It was just high enough to be both convenient to water and safe from flooding. It was near the confluence of two branches of Harshaw Creek. The flood plain that was now a series of small ranches would have been a perfect place for seasonal floodplain farming. This place was special, in part, because of its geography.

It was also special because of its ecology: the dominant plants on the hill-side—mesquite, oak, manzanita, agave—are all edible. Mesquite beans were ground into flour, and acorns were roasted or leeched and ground into meal. Manzanita berries can be enjoyed like raisins; and agave is roasted and then fermented into mezcal and tequila. But along with being a supermarket, this place was also a hardware store. Branches from willow saplings growing in the flood plain were bent into the structural supports of homes. These pit house walls were covered with the sacatone grass, folded and sewn into mats.

This particular place was a remarkable spot, but in every single place a story will unfold. First you need to notice its details; then you need to learn more about what you see. Eventually the pieces will reveal the narrative of your chosen place. Once the story is clear, you can create a treasure hunt that shares these clues—this embedded knowledge—with others.

CONSIDERING YOUR PLACE

As we begin to create treasure hunts for others, we should consider the tremendous wealth waiting to be found in our communities. We can easily make a treasure hunt at random and hide a little box out in the landscape, but better still is a Quest that takes good advantage of the subtle clues lurking in our communities, the stories unfolding right before us. *Here* is where you can see the storm track of the "Hurricane of '38"; *over there* you can follow the glacial erratics and till from the last ice age; *that* was the home of Hetty Green, once upon a time the richest woman in America. To make a Quest like that, however, we have to breathe in the spirit of a place.

This period of inspiration—this breathing in of the spirit—does not necessarily take a long time. In fact, it is something we can begin to do every single day as we commute to work or go about our daily lives. Every day we pass through a world that is literally buzzing with life and bursting with ripe details. Yet how much do we take in—really *see*—along our way? Can we train ourselves to separate the wheat from the chaff, and see the keys that unlock the hidden stories—and hearts—of our communities?

How aware are you of your surroundings, after all? For what or whom is your community named? From where you are sitting, which direction is north? What is the current phase of the moon? If you were a raccoon, where in your neighborhood would you choose to spend the night? Why? If there were no stores, where would you forage—and what might you find to sustain you? Where did the first residents of this community live? Why did they choose to stop and stay here? Where did they get their drinking water? What did they eat in summer or winter? Asking questions like this helps you open your eyes, your gaze, just a bit wider.

And why the details? Aren't our brains—and our desks, too—already covered with too many lost, forgotten, or torn fragments? Why bother with more clutter? Because folded in with these details are the timeless stories our places tell. "The universe in a grain of sand, eternity in an hour," wrote William Blake. He was right: if we miss the details, we miss eternity too.

Your Daily Routine

Your daily excursions through your community present an excellent opportunity for noticing and considering the clues that beckon from both the natural and the built landscapes. Let's consider Steve's commute to the office as an example. Going to work, he turns out of his driveway onto a dirt road, following the white rush of Barker Brook south toward Vermont Route 113. This road curves back and forth three or four times before a gradual, sloping ascent into the small hamlet of Thetford Center. Just before arriving there, he passes a softball field with a wooden backstop on the left and a prominent brick mansion on a knoll off to the right.

If he looks closely, he can see a plate adjacent to the mansion's door that reads "1822." Driving on, he quickly passes a dozen buildings—half of them brick, plus a few wooden capes—and then begins the steep ascent up and over Thetford Hill toward the interstate.

That's Thetford Center, 05075: a combination village store/post office/gas station; a church; the town office; the Community Association building, where

3.3 The Hezekiah Porter house (1822), Thetford Center, Vermont. Photograph by Simon Brooks. This lovely brick Federal-style house rests on a knoll overlooking the village of Thetford Center. But why is this house here and not somewhere else? Who built it? What were they doing here? Buildings are not only objects—they are also clues.

Thetford residents vote; two cemeteries; a garage; Elmer Brown's plant nursery; and a sprinkling of houses. Not much. Yet there are a number of stories hidden even here. Let's consider a few of them.

First of all: Why is this *place*, Thetford Center, here at all? Just beyond the "1822" mansion, we notice that the land drops off into a narrow river valley. There, straddling the Ompompanoosuc River, sits the Thetford Center covered bridge. Sticking our noses out of the south side of the bridge, we can see that it traverses a number of small falls. For the first 150 years of this town's existence, these falls powered a series of mills. Free energy—the falling water—was a key element that gave rise to this particular settlement of Thetford Center, Vermont.

No matter where you live, your community, too, rose out of particular qualities in the landscape. Communities have always needed energy, water, and food. There are many conditions that might spawn a community: a harbor, a river, waterfalls, rich soil, a mineral deposit, even something as simple as free or cheap land. Places also arose out of the forces of settlement and migration: the path of railroads, the "King's Highway," or the Erie Canal, to name a few.

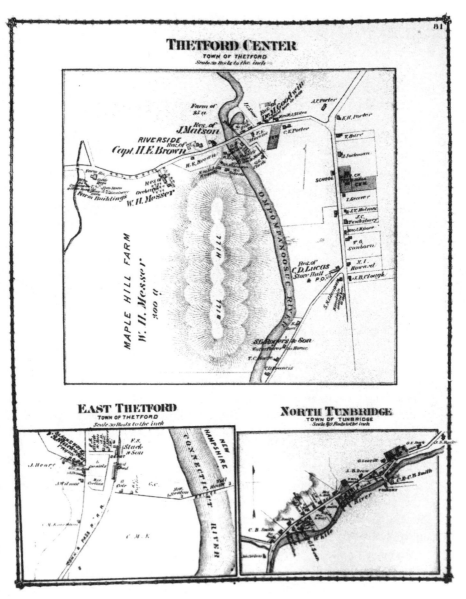

THETFORD CENTER
TOWN OF THETFORD
Scale 30 Rods to the inch

EAST THETFORD
TOWN OF THETFORD
Scale 30 Rods to the inch

NORTH TUNBRIDGE
TOWN OF TUNBRIDGE
Scale 40 Rods to the inch

3.4 Beers Atlas Map of Thetford Center, 1873. Photograph by Simon Brooks. A 130-year-old map of Thetford prominently displays the Porter house and a series of mills along the banks of the Ompompanoosuc River. Antique maps residing in your community can help you deepen your understanding of your Quest site. *Map courtesy of the Thetford Historical Society.*

Try to see the patterns of your own particular place that are still evident in the patterns of the natural landscape: hills, valleys, plains, river confluences, drinking water sources, strategic overlooks.

Now let's turn our attention back to the 1822 mansion itself. Why this brick mansion on the corner? It turns out that this was once the home of Hezekiah Porter, a mill owner. But Porter also owned a brickyard across the street from his house, where the softball field now sits. This brickyard was the source of the red fired clay that is the key ingredient in Porter's mansion as well as the Timothy Frost Methodist Church and other town buildings.

Let's look at the white and green Thetford Center Community Association, just across from the Village Store. Its bell tower—and the line of tall windows marching along one side—gives this building's identity away. This was once a one-room schoolhouse. Until consolidation of the school district and the building of the new elementary school in 1961, this was the Thetford Center School.

Why would such a tiny settlement need a church and a school? It turns out that nearly every small settlement in the 1800s had a church, a school, and a cemetery. When the main mode of transportation was traveling on foot or by horse, it was difficult to get very far (especially in winter). Making *here* work was easier and more preferable then heading *there*.

The patterns of your own particular place, too, will still be evident in the patterns of the built landscape: look at the railroad beds, power lines, canals, roads, older buildings, plantings, parks, and monuments.

Recalling the falls and mills, it becomes clear that once upon a time Thetford Center was a hub of commerce. It was the place where the jobs were, and it boasted a grist mill, a cloth mill, a trip-hammer shop, a shingle mill, a sash and blind factory, a cobbler, and a wheelwright. This is hard to believe today, as folks speed along in the line of cars commuting to jobs in Lebanon and Hanover, New Hampshire, fifteen miles away. But looking out the south side of the covered bridge, we can still see the foundations, stepping like forgotten stone stairs down the falls toward the Connecticut River. Each step tells a hidden story of this community.

You can pass all of this in a flash on your commute and never take any notice of this place at all. But once you connect with even a single detail, the Quest begins. Details become clues. Clues are seductive, and along with them arises the impulse to discover. You pull over and get out of your car or walk around a neighborhood where you've never bothered to stop before. Perhaps you talk to a person you've never met and ask a question that leads to a conversation. Or perhaps you simply linger, picking an apple off a tree as you peek in the windows of the old schoolhouse.

Beginning to notice the entwined patterns of landscape and settlement will enrich both the depth and the quality of your Quest. This is an easy thing to begin to do, and it is a process of investigation that can last for the rest of your days.

Your Place Tells a Story

But what if you live in the suburbs, a place seemingly newer—or more crowded? What can you do then? Are there still details and stories waiting to be found? Before moving to Thetford Center, Steve lived in Lafayette, Colorado. Lafayette is out on the plains, a half-dozen miles east of Boulder. Presently, Lafayette is home to supermarkets, department stores and chain stores lined up end-to-end in strip malls, but let's ask the same question of it: why does Lafayette, Colorado, exist?

If you look for it, back behind the K-Mart you will find it: the old downtown, sporting Mexican restaurants and antique dealers. Main Street comes to a T, dead-ending into the elementary school and a recreation center. To the east and west of Main Street lies a fifty-square-block neighborhood of houses a hundred years old or more. A simple observation: these houses are all approximately the same size; they all share a north-south orientation, and they all have similarly sized lots, save for a few very notable "big" house exceptions, which also share a common pattern: these are on corners and what appear to be "double lots."

What does this settlement pattern show us? It shows us that the lots were plotted out all at once, and the houses were perhaps built in the same period, too. Which leads to a deeper question: why?

If you buy a house in Lafayette (or examine a deed), the story becomes clearer. You (or the bank) may own the home—but someone else owns the coal, the natural gas, and any oil that may be beneath your little green lawn. And buyers beware: the house is just a few feet above thousands of feet of tunnels that literally honeycomb the plains. Why? Because Lafayette was a coal town. These tunnels and shafts are hidden remnants of the old coal mines. Lafayette miners dug the coal that powered the region. If you live in "old town" Lafayette, your house, most likely, was a miner's house . . . unless, of course, you are in one of the big, corner-lot houses. If that is the case, then congratulations: you are the proud owner of the former mansion of a mine owner.

The story of Lafayette, like the story of Thetford Center, is a story of energy. While in Thetford the clues were falling water and a brick house, in Lafayette the clues are hidden underground, but also in the street signs and even the city's name. Lafayette, Colorado, is named for Lafayette Miller, who died in 1878. His widow, Mary Miller, with the help of John Simpson, dug the shaft of the first coal mine, the Simpson Mine, back in 1887. The "old town" was laid out in 1888, and the first homes rose quickly on the plains. A second mine, the Cannon, opened shortly thereafter. Walking south from the recreation center toward the Wal-Mart, you can feel the past come alive in Lafayette's street signs: Simpson Street, Cannon Street, and so on.

No matter where you live, you will find history embedded in place names: the name of your town or city; street names; river names; the names of hills.

What if you live in an urban neighborhood? Or in what seems like an endless suburb? Are there stories hidden there too? Yes! Every single place has a history, characters, natural communities, and stories.

To learn to see these traces, first look at what is already staring you in the face: note what is present and also what seems absent. The things missing are the ghosts. Look for uniformity. Are all the buildings of the same design? The same materials? The same era? And look for anomalies: a lone farmhouse among strip malls, an incongruent building design, a pod of attached industrial or agricultural buildings. Pay attention to geographical features like ditches, ravines, and hills. Cultural artifacts, too: the nameplates on buildings, cornerstone dates, historical markers, street signs, the old photographs you often see in banks.

Steve graduated from Nova High School in Fort Lauderdale, Florida. Back in the 1970s, Nova was a new, commuter magnet school set in the middle of a field. Definitely "nowhere," as author James Howard Kunstler might have called it in *The Geography of Nowhere.* An instant, brand new school set on a giant, vacant lot—just where was the story in that prefab place?

In gym the kids used to play kickball out behind Nova 2, just beyond the asphalt basketball courts. Chasing after the ball in the outfield they noticed that there was a strip of old concrete, seeding in with grasses, that shot across the playground. The strip eventually slipped underneath a chain link fence and disappeared into the burr fields beyond. While playing kickball, if you caught the red rubber ball with a good swift boot that sent it to the concrete strip, it would roll on and on. A ground ball transformed into a home run! Later, in high school, the kids would walk along this crumbling path while taking a shortcut to McDonalds, Burger King, or Wendy's at lunchtime. But just what was that old, concrete strip?

Have you ever heard of the "Devil's Triangle"—also known as the "Bermuda Triangle?" The Devil's Triangle is the name for a geographic place, an area of land and sea stretching roughly from South Florida east to Bermuda and then south toward Cuba. If you take those three locations as corners, the space—or place—between them is the Devil's Triangle.

The Devil's Triangle, thanks to the efforts of a few sensationalist authors, has achieved notoriety for being a place where planes, boats, and people disappear. They all vanish "without a trace." The merchant ship *Marine Sulphur Queen* vanished there; a steamship with close to four hundred passengers at the dawn of the twentieth century disappeared as well. At mid-century a convoy of five United States Army bombers disappeared. *Perhaps this is the very site where Atlantis vanished as well,* the authors tell us.

From where did these five bombers take off? Fort Lauderdale Airforce Base. And where was Fort Lauderdale Airforce Base? The Nova Schools were built on the site of the Fort Lauderdale Airforce Base! That old strip of concrete was a runway; this very place is the place where those five bombers took off. A hidden story lurking on the Nova campus is the fact that it marks one of the three corners of the Devil's Triangle.

How Many Paths?

By paying attention on your community excursions, with just a little additional time given to spontaneous side investigation, the environment surrounding you—your place—will start to come alive. By building just ten or so minutes more into our commute times, we find that we can choose among eight or ten different routes to work. What point does this additional traveling time serve? By mixing up our routines, and paying attention, we can come into deeper touch with our sense of place. As the weeds die back in the late fall, a new cellar hole is revealed. Driving home with open windows on a spring evening, we can hear the frogs singing and learn of the vernal pools that line the height of Union Village Road between Union Village and Norwich. A brief stop spent listening and observing teaches us the songs of the wood frog (with its brown mask and deep ducklike quack), the spring peeper (with a cross on its back and a high-pitched "peep"), and the American toad (with its bumpy back and sweet and musical "trrrilll"). Later on, we learn that each frog's raucous song betrays the salamander's silence. Each one of these details helps us become more in tune to and aligned with the stories continuously unfolding around us in this place, our community.

By mixing up your daily rambles, you just might find out:

➢ how your community got its name;
➢ who its earliest residents were;

- where and when it was first settled by Europeans;
- what community features were most important to survival in the nineteenth century, the twentieth century, and today;
- the names of early and important families;
- your community's industrial heritage;
- the native grasses, shrubs, and trees;
- resident bird and mammal populations; and
- animal migration corridors.

The first step in Questing is paying more attention to the world around you. When you will be working with groups, find the time to explore a bit on your own before beginning your Quest project. You will find it much easier to engage the group afterward.

DIGGING A LITTLE BIT DEEPER

A concurrent step in your preparation is to dig deeper. As you begin your Quest project, it is helpful to search for the following things.

Written History. See if there is a written history for your town, county, city, or neighborhood. If so, get a copy and begin to skim through it. You can read it cover to cover, or you might try this: open the book at random and read in snatches until you find a storyline that captivates you. Now that you are eager to discover more about *that* particular narrative, you can start back at the beginning—while simultaneously discovering and plowing through as many other parallel sources as you can find.

Contemporary Maps. Find a contemporary map or USGS topographical map of your community. Look for the areas that are shaded in green—the natural or open spaces—or the watercourses, or anything else that stands out and surprises you. Try to visit a few of these sites.

Historical Maps. Find at least one historical map of your town, county, city, or neighborhood. By comparing an old map with a new one, you are forced into a creative tension between what once was (or seemed so) and what now is (or appears to be). This tension is not only educational but also inspirational, serving as a further stimulus toward inquiry and toward taking the long view. Just how far have things come? Why? Where does this place, or this community, seem to be going? We must consider how different values and choices inform community decisions and affect our places over the long haul. Old maps also reveal compelling features like mines, mills, quarries, brickyards, one-room schoolhouses, poor farms, factories, businesses, and much more.

Local Museums. See if there is a local museum or historical society. Arrange for a visit. Ask for a general tour focusing on the different primary and secondary sources housed there. This will help you to discover more quickly some of the key stories, natural and cultural, of your community. Your visit will also serve as a reconnaissance mission regarding materials that can offer your students or group the opportunity for first-hand discovery and learning experiences.

Neighbors. Introduce yourself to someone who has been a resident of your community for a long time. Ask this person to walk a place with you, sharing the sights, their memories, the shadows, and the "ghosts." Lovers of the land are a generous lot—and usually more than happy to share their wisdom and memories. You may find yourself tagging along on a bird survey of a neighborhood park in spring, or sloshing through the muck, thigh-deep in your neighbor's mud. Both can be fun.

Nature Centers. Visit your closest nature center to learn more about your region's native flora and fauna. Go on walks sponsored by the Audubon Society or similar organizations. Carry a field guide to plants and wildlife along with you in your travels. If you can spare the time, stop, look, and then look things up. More important than knowing all of the names is developing your familiarity.

Species a Day. Build your knowledge of the others who share your place by learning a "species a day." Every day (or week), try to learn the defining characteristics of one more neighbor. Insect, frog, weed, tree—it's up to you. At first, it's not even important that you get the name of whatever it is right. Just begin to recognize this new friend's structure, proportions, and defining qualities.

Some people find that it's easiest to learn to see by trying to draw. Sketching makes you look a little longer and requires you to look back and forth from the subject to your image, again and again. As you create your composition in parts, you learn to see how these fit together and function as a whole. It doesn't matter if your drawing is accurate. You are developing an intimacy with your surroundings.

After you have the basics of structure—"Ah, that's a papaya tree"—you

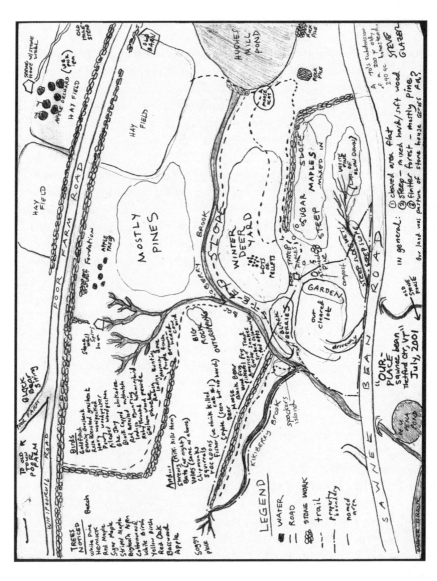

3.5 Sense of place map. Drawing by Steve Glazer. A sense of place map joins words, images, and personal experiences to create a visual record of your place. This record is a starting point for identifying possible themes for

can expand your ring of perception to include the context: not just your object of focus but where you are seeing it. Once you recognize a number of individuals and their contexts, you will begin to see the patterns and processes that make up the larger landscape.

Sense of Place Map. You can also set your mind freely roving through the special places of your community by creating a free-form map of the places and ways you connect to it—a solo brainstorm. Find twenty minutes, a piece of paper, and something to write with. If you want to really open up, try to find a big piece of paper and some colorful pens or pencils.

Now draw a "map" of your home community. It might look like a geographic map, or it might be more representational—a sun, or a tree with roots and leaves representing different aspects of your life and your relationship to your community. Next, place the following symbols on your map. You may choose as many as you like from each category, and you might want to make notes along with the symbols, to identify them:

> a favorite place in your community to go for a walk;
> a favorite public place to chat;
> a local sacred place or healing place that always makes you feel better;
> an older person in the community whom you appreciate knowing and spending time with;
> something special you've noticed in your community that you suspect few people know about;
> a place where you connect personally with an element of your community's local or regional economy; or
> a place where you connect personally with the human history of your community.

Or choose other categories and elements that interest you.

A SPIRIT DANCES

In Lyme, New Hampshire, we wanted to make a Quest focusing on the small village of Lyme Center. As we began working on the Quest, it quickly became evident that there was an invisible, powerful spirit lurking right there in the center of town. While we all could see the tall, white Academy building proudly standing there, to Lyme Center's older residents this building was forever the backdrop for the biggest show in town—the square dances that were once the center of the community's social life.

Walking through the village with Ken Elder, a resident of Lyme for more than eighty years, we saw each building quickly transform from a structure into a home. Each home had a family, with faces of children, parents, grandparents, great grandparents. Everywhere, there were memories. The walk with Ken led to an oral history collecting event called "Remembering the Lyme Dances," because other community elders remembered them, too. These dances were the basis for courtship, marriage and divorce, fights and friendships. "Remembering the Lyme Dances," an event held as part of the community's Old Home Day, featured readings from the collected oral history, spontaneous offerings of memories, and real live square dancing with two old-time callers. The memories of more than a half-dozen seniors were collected, transcribed, and then edited into verse to become the Lyme Center Dances Quest, which turned the hidden spirit of a place into a public journey.

The Lyme Center Dances Quest
A Walk through Lyme Center's Past

As you walk down Acorn Hill, please pretend that there's snow.
Though it's a dark winter night there's somewhere special we'll go.
It's the 1940s—and no, you are not traveling by car—
But walking to the Academy, where the dances are.

You'll be gathered together in groups of two, perhaps three;
And carrying kerosene lanterns, which at night help you see.
And each house that you pass? Why you know it by heart!
It belongs to the Reeds, the Cuttings, or to Charlotte LaMott.

At the bottom of the hill, that's Grandpa Cutting's place.
It's the building right across the street you now face.
Take a right turn, watching for traffic as you walk,
And of the buildings you pass I'll continue to talk.

On the right-hand side, why that's Charlotte's, for sure.
And across from there, that's the old grocery store.
Pearl Dimick's house is next door, beyond Charlotte's place.
And looking across from there, Sanborn's Sawmill you face.

Now head up the driveway to the Lyme Center Academy.
While now sometimes empty, it was the heart of the community.

People gathered here on many a Saturday evening
And had a great time 'til by lantern they'd be leaving.

Walk to the right (east) side of the building:

Back then, outside of the building you'd walk up some stairs
And Fred Welch, the ticket-taker, he'd meet you right there.
It would cost you 35 cents to get in to dance.
You'd pay your money and take your chance.

Laura DeGoosh says:
"My dad ran the dances at Academy hall.
Everyone came—and we all had a ball.
There was violin, piano, and banjo, too,
And steamed hot dogs there in the corner for you."

Now walk to the back of the Academy, and look up to the second floor:

Roger Rich tells us:
"Music was usually served up by local talent—
Charles Balch, Ken & Ray Uline, and Kenneth Dimick.
Often, Alberta Tupper called the dances,
Sometimes fifty couples 'round the room a-prancing."

Albert Pushee:
"Alberta Tupper, she would use a megaphone to call,
That way you could hear clear across the dance hall."

Ken Elder:
"The Uncle to all the Cuttings out here, he played fiddle.
And so did Frank Humiston and Charlie Canfield.

"I didn't go to the dances until I was sixteen.
Emerette Pushee, she taught me some things.
She was much older—as old as my mother,
But she would grab my hand and then teach me another."

Continue to walk counterclockwise around the building:

Ken Uline:
"One time the band was nearly all my relations—
myself, my brother, and brother-in-law; my sister calling changes.

The Spirit of a Place ➤ 43

And whole families would turn out for the show . . .
Sometimes with a jug of hard cider in tow!"

Mertie Balch:
"We'd dance to 'Soldier's Joy,' and the 'Virginia Reel.'
Also 'Lady of the Lake' and the 'Wabash Cannonball.'"

Charlotte LaMott tells us:
"Dora danced with Fred Welch; while Mertie would dance with me.
They weren't enough boys—so girls danced together, you see.

"There were police officers, too: Francis LaBounty, Harry Franklin.
Now Harry, he wouldn't let the young folks have so much fun!"

While Charlotte's father said: "Dancing was a waste of good shoe leather,"
The Lyme dances were a wonderful way for people to spend time together.

Ken Elder:
"Back then, gasoline was rationed. And just where were you going to go?"
Here in Lyme Center, on Saturday night, the Dance was the big show.

For the last part of your Quest, look near the entry for the treasure box.
Hidden inside, you'll find the faces of all the people you've heard talk.

As you remember these faces and stories you can surely see
What it means to be a member of a living community.

After you've looked through the box, please sign in . . .[1]
And we welcome you to come visit Lyme Center again.

No matter where you live, there is richness and vitality lurking just beneath what perhaps appear to be the ordinary surfaces of the world: buildings, fields, roads, rocks, bridges, and brooks. A little wide-eyed digging reveals that there are truly veins of gold—and the riches of our communities can be tapped into, remembered, shared, and cherished.

4) Choosing a Quest Route and Site

SO, WHERE WOULD YOU LIKE YOUR QUEST TO GO? Where should it start? What route would you like it to follow, and do you have a final destination in mind? Simple questions, yes—but sometimes the answers are a bit more complex.

There are practical considerations as well. Is there an underutilized public space in your community that would benefit from more traffic? Conversely, is there a well-loved and well-visited place that does not need additional visitors? Is there a spot that is magical but perhaps a little unsafe? Is there a spot that is the centerpoint of a local story so compelling it's just begging to be told? Is that spot open to the public and easily accessible by mass transportation or from available parking? The process of choosing your Quest site offers an opportunity to learn more about the nooks and crannies of your community that is richer than you ever dreamed.

GENERATING IDEAS

Community Mapping

A good way to generate a lot of site leads in a short time with your group is to undertake a community mapping exercise. This exercise works well with up to twelve goodnatured people. Gather around a large sheet paper, the bigger the better, and distribute markers to everyone. Draw a shape roughly approximating the shape of your community in the center, with lots of room around the edges. Ask your group members, "What's important to have on a map of our community? What places do we know about that we think other people would find interesting?" Invite them to sketch in and label anything that seems important, for instance:

➤ key public buildings like the post office or a museum;
➤ key private buildings like a popular store or well-known restaurant;
➤ geologic features like hills or valleys;
➤ watercourses like rivers, ponds, or lakes;
➤ cultural landmarks like statues or historic markers;

- public land like municipal parks or surrounding national forest;
- bordering communities;
- natural features like a favorite tree or swimming hole;
- their favorite places.

In this somewhat chaotic but creative atmosphere, ideas breed more ideas, and you'll find that your map is soon overflowing. Remember to notice the people who seem to know everything—you may need them later in the project. Following the exercise, post the map in a place where group members can continue to access the ideas generated, or add more.

Convivial Research

If you are working with a larger group of community members (twenty to sixty or more) and have the time to lead them in a more structured activity, try "Convivial Research," named for the fun and social atmosphere in which it is carried out.

In preparation, consider the questions that you would find most interesting and helpful to pose to a group of fellow community members. Develop that list of questions, making it long enough to provide one question for every two people in the group you will be working with. Assuming a group of up to sixty, we have provided a list of thirty sample questions below. Use them as is, or tailor them to your community.

- Your cousin who is an artist is coming to town to make a documentary video about arts and crafts in our community. Who would you want to introduce her to, to get started?
- In your opinion, what are the three most important or distinctive buildings in our community?
- If you were designing a new logo for our community, what key stories or themes in the life of our community would you want it to depict?
- In your opinion, what are the three most important or distinctive geographical features in our community?
- How does the story of our community connect to the "bigger stories" of the region, or the nation, or the world?
- What are the two most fun buildings in our community, either inside or out?
- Where do you like to go in our community to look at water?
- If you could look into a crystal ball and see the year 2030, what are two things you would hope to see had *not* changed in our community?
- If you could look into a crystal ball and see the year 2030, what are two things that you would hope to see had changed in our community?

- What three organizations in our community might be interested in cosponsoring a Quest?
- Do you know of any unusual sources for interesting facts or photos about our community's heritage (primary or secondary source material)?
- If you could bring a small group of children to spend an afternoon talking with one of our community elders, who would you choose? Why?
- What do you consider to be three important historical events in the life of our community?
- Who do you think knows more about the history of our community than anyone else?
- Who do you think knows more about local natural history in our community than anyone else?
- Your uncle, who is a history buff, is going to be visiting next week. Which three cultural landmarks will you show him, to impress him about our community?
- What public land has been protected in our community—for example, parks or wild lands? Is there more land that should be protected, and if so, which parcels?
- Think about the borders of our community, where it meets the next community (or the river or ocean). Which two of these edges do you find most interesting, and why?
- What is the most interesting natural feature that you can think of in our community—for example, a pond, a deer habitat, or a huge old tree?
- If there is one place in our community where it is most possible that magic might happen, where is it?
- What is your favorite place in our community to go for a walk or a hike alone?
- Where do you go in our community where you almost always bump into people you know?
- When you're feeling blue, is there a place you go in our community that sometimes inspires you or makes you feel better?
- Is there some little detail you've noticed in our community that you think few other people have noticed? If you're willing to reveal it, what is it?
- What is the most sacred place in our community, to you?
- Think about the "underbelly" of our community, the dark or mysterious or vacant places. Are there any of these that would be safe and appropriate to open up for a little Quest exploration?
- What two places would you bring a group of middle school students to if you wanted to demonstrate to them the underlying driving forces of the economy of our community?

➤ You have decided to take up photography. What three unusual views or beautiful places in our community will you use as your first subjects?

➤ What is your favorite place in our community, outside of your home?

➤ How do people in our community find out about public events and recreational opportunities?

Next, prepare a "Community Research Response Form" that includes instructions, a space for the individual questions, and a two-column grid in which there is a narrow column labeled "names" and a wider column labeled "answers." If you have a very large group, you will probably want to prepare a second page of the grid as well. Print enough forms so that there is one form for every two members of the group. Glue or write a different question in the space on each form. The more colorful and decorative you make the forms, the more convivial this exercise will seem—which will help to get the creative juices flowing.

Gather the group and explain that they are going to be conducting important research about the special features of your community. You have arranged an opportunity for them to interview notable experts on the subject—each other! Following the interviews, they will be collating and interpreting their data to present to the rest of the group.

Break group members into pairs and assign each pair *one* question. Instruct them not to answer their own question but rather, working as a pair, to circulate through the room, approaching all of the other pairs and asking each member to succinctly answer the question. They should then answer other groups' questions in return. Encourage the groups to keep moving, spending no more than a few minutes with each group/question. All of the groups may not get all the way around the room, but they should hit as many other groups as they can, taking brief but clear notes on each answer. Try to allow at least twenty to thirty minutes for these interviews.

At this point, you can choose to wrap up the exercise in one of three ways:

1. If you want to build a sense of community and commitment to the project, and you have forty-five minutes, supply each pair with a large sheet of paper and some markers or crayons. Instruct them to take about twenty minutes to create a poster that graphically depicts their findings, making sure to include their question on the poster. Tape the posters up around the walls of the room as they are completed and lead a "gallery tour." Let each group carousel around to visit all of the posters, studying them and raising questions. Ask each pair to briefly introduce two highlights of their findings.

Researching Communities

- Working with a partner, circulate through the group, find another pair and ask each person this question. You should take about one minute per interview, so remind your interviewees to keep their answers succinct. You can keep track of who you've interviewed and of their answers on this sheet. Note: "Community" means whatever local area makes sense to each interviewee.

Your Question:

What is your main source of local news in your community?

Your Interviewees' Names	Their answers	Free newsltr:
Melissa	News paper / Keene Sent. /	TROY news
David	" / Malden Observer	
Joanne Buckner	Free paper Daily Sun	
Diane Bjorkman	Free paper – Malden Observer	
Morgan	Channel 2 – News station → local	
Brian Hall	Valley News – low cost	
Irene	Keene Sent. Biased IMHO	
Wendy	Channel 2 / local channel	
Carol/Karen	Word of Mouth	
Nancy	Local Paper Keene Sent. Very Good IMHO	
Delia	Counterman at Charlies Store / Parent gossip at P.U. / D.O. at school	
David	Local Grapevine / local store	
Amy	Keene Sent. Good Paper IM HO	
Hope	WEFCR / NPR Radio + Local Papers (	
Ariel	Television / Malden Access	

4.1 "Convivial Research Response Forms" focus on specific questions and are used in the Convivial Research process. Conducting group interviews is a fun way to gather a lot of information in a short time.

2. If you have less time, conclude the interviews by holding up a newspaper and saying that it is an edition of your local paper from five years in the future. The lead story on the front page, above the fold, is related to their particular question. What does the headline read? Give each pair a couple of minutes to develop a headline, then sweep around the room, asking groups to read their headlines.

3. You can also ask for a simple report. Give each pair a moment to read over their findings, then ask them to share something that surprised them from doing this exercise.

No matter which concluding exercise you choose, be sure to gather the sheets to serve as a data bank. You may want to type the notes and distribute them to each member of your group, so each participant has a copy and can continue to add on to it. Or you might find a partner group to edit and produce this community resource directory.

FACTORS IN SELECTING YOUR QUEST SITE AND ROUTE

We have found that there are a few key considerations that may affect your choice of a site and route for your Quest.

Connecting Site and Story

Place and story are inextricably interwoven. Sometimes a story is so compelling that it is the central factor in the selection of a site; other times the place comes first. In either case, the ways you link the two will impact the power of your Quest. Does the Anthony Mangione Quest start at the beautiful Greek revival library that he built and end in the cemetery where he is buried, or vice versa? Which has better parking? Which has the best treasure box hiding place? Which direction will allow you to unfold his story best? Should the Green River Water Quality Quest include a stop at the sewage treatment plant or stick with stops along the river? How central is the sewage plant to the story of the clean water campaign? Is it a centerpiece or a distraction? Is the public allowed near the treatment plant after hours and on weekends?

Choosing a Starting Point and an Ending Point

Every Quest has a beginning, middle, and end. Before you get too far along with your Quest, you will need to choose a natural starting and ending point.

4.2 Good Quests join site and story in a seamless whole. Photograph by Jon Gilbert Fox.

Out in the country, this might mean a pull-off beside a gravel road; in town, the front steps of a prominent building. Consider a wide range of transportation options in selecting your Quest's starting point. Wherever possible, we encourage you to include spots that are walkable from the community center or accessible by mass transportation. If your site is way off of the main transit routes or in a rural area, you will want to be sure that there is adequate, safe, public parking. We have found that even the most civic-minded citizens tire of blocked driveways and wheel-rutted lawns, so assume that your Quest will be popular and protect its neighbors. A key component of a finished Quest is the set of directions to the starting point. Remember that these directions must be precise, for example, "To get there: Take I-91 to Exit 13. Travel west into downtown Norwich. Park near the bandstand gazebo on the Norwich Green across from Elm Street."

Clustering Quests

As your Quest program grows, you may want to consider clustering multiple Quests within a single area, perhaps even designing several to depart in different directions from the same central point. This will allow your Questing audience to enjoy several Quests in one outing, without having to climb into their cars or hop on a bus. This also facilitates school and community groups' use of your Quests as a fieldtrip activity. By dividing the group into smaller subgroups, people can head off in different directions on different Quests, allowing each individual to become more engaged. This is preferable to twenty people clambering for the same clue in the same place at the same time.

Offering Targeted Experiences

Will you consider your Quest program to be a complete success if you find entries in the treasure box log that read "A wonderful place to visit again and again!" or entries that read "Yes!! We solved it!" or entries that read "Fascinating story—I had never realized . . ."? Some Quests introduce people to a pleasant walk that they will want to repeat often, bringing along friends and family members. The Pinnacle Hill Quest is such a walk, and sign-in book entries indicate that there are many repeat visitors. Visitors only solve the riddles once, but they come back to enjoy the walk and sign in each time. Other Quests create a fine mystery that people enjoy puzzling over until they jubilantly find the solution, then they go on to the next Quest, unlikely to return. Other Quests accomplish the goal of telling the stories of a community. All of these experiences make for good Quests, and you may want to have some of each in your program.

Accessibility

Key factors to consider in accessibility include whether key spots along the Quest route will be open when visitors get there and whether Questers will be able to manage the route's terrain.

In our part of northern New England, we have found that it works well to establish a Questing season—say, Earth Day (April 22) through Thanksgiving. You may choose to extend the season to year round if the climate in your area permits. You might also wish to include indoor Quests. The key with Quests that either end indoors or utilize indoor elements, however, is to check and clearly note the hours that the Quest is open to the public. This goes for public parks and other settings with limited hours as well.

The idea of Questing appeals to people with a wide range of physical interests and capabilities. We have found that it works well to design a range of options, including Quests that use:

➤ smoothly paved sidewalks or paths and buildings that are wheelchair and stroller accessible;

➤ bumpier sidewalks and paths with curbs or a few stairs;

- rough footpaths or hiking trails;
- bike paths, which are often also accessible by roller blade and skateboard;
- waterways accessible by canoe, kayak, rowboat, or sailboat; or
- cross-country ski trails.

The trick is to find a way to clearly state the level of physical difficulty at the outset so that Questers can plan accordingly. Up here in ski country, one group chose to rank their Quests according to the commonly used skier's difficulty code: green circle for beginners, blue square for intermediates, black diamond for experts (and double black diamond for "you've gotta be crazy!").

Private or Public?

In general, we have found that it works best to design Quests on public property—along public streets, in public parks, in public libraries, and the like. We have also had good luck designing Quests that include stops on private properties that are open to the public, such as museums, stores, and nature centers. It's trickier to route a Quest through a private property that is not normally open to the public, but it is not impossible. It relies upon a civic-minded property owner who has a clear sense of what is involved and freely gives written permission for this specific use. For more on permissions, see below.

Safety and Security

Obviously, as with any public program, it's important that you consider the safety of the people you attract to your Quest. If you've labeled your Quest as being physically rugged, then it is legitimate for you to include some rocky scrambles or slippery logs. Choose routes that stay away from the edges of waterfalls, rickety wooden ladders, lots of broken glass, and the like, as well as places where personal security might be a problem or where there is a high likelihood of vandalism. If you hide your box in a dark rocky crevice where snakes or stinging insects may be present, include a warning in your final clue to note this possibility. Learn to identify poison ivy, stinging nettles, and other troubling vegetation so you can avoid placing your box in the middle of a patch of the stuff.

Impact

In addition to the safety of the Questing public, it's important for you to consider ways to protect the natural and cultural resources of your site. This might involve routing people around areas of delicate vegetation through the use of humorous clues that admonish them to stay on the paths; or it might involve

4.3 Consider the future impact of a group of twenty tramping toward your treasure box. Photograph by James E. Sheridan.

reminding them not to talk loudly or touch things in museums. Whether it involves private or public property, indoors or out, be sure to plan your Quest so that it minimizes any impact and respects the gift of access, as indicated in this excerpt from the Jonathan Wyman Sawmill Quest:

> After you park, take a quick look
> Where a fall of water becomes a brook.
> It's privately owned, so to do what's best,
> Just eyes should admire this part of the Quest.

PITFALLS

We'd like to offer you the benefit of our experience so that you can avoid some of the pitfalls we learned about the hard way!

Seasonal Changes

When planning your route and end site, be mindful that in some regions vegetation changes dramatically throughout the seasonal cycle. "Walk a little bit more, you're doing fine. Then look far to the 'Dead End' sign" seemed like a great clue in April, but by July the sign was nearly invisible behind thick green foliage. Bare ground in November or April may have three-foot high ferns or blackberry bushes in August and a dry arroyo in fall may be a rushing torrent in March, making for not only confusing clues but safety risks. This goes, too, for hiding your box—a nice secret spot tucked against a rock in thick grass will be quite visible once the grass dies off.

Staff Changes

Some wonderful Quests rely on interaction with store clerks, concierges, librarians, and other local characters. Other Quests rely on a staff understanding that the "N" posted in the window has to remain visible no matter how the window display is modified. One Quest asked Questers to walk up to the person behind the cash register at the final destination and say, "Hello. I'm ill. I need some sarsparilla," at which point the person was supposed to produce the Quest box. You will need to work with local business owners and supervisors to flesh out such arrangements, and to be sure that they remember to inform their new employees about them.

Changes in Appearance

There's nothing like a coat of paint to spruce things up—or to mess things up, if you're talking about Quest clues. "Pass the statue with the torch. Forge ahead to the blue front porch": need we say more? While you can't control whether people in your community choose to repaint, relandscape, rename their stores, or erect new signs, you can keep an eye on these changes and update your Quests accordingly. The best practice is to base your clues—as much as possible—on the features and elements that are least likely to change. It is better to rely on the number of pillars on a porch than its color.

Fragile Places

Some places are just too fragile to support increased visitation. We know of a beautiful bog, for example, that would make a magical Quest site, but we feel it would be irresponsible to choose it, as it doesn't have the boardwalks or infrastructure to keep it from getting wrecked by a lot of foot traffic. Remember that there are many secret places in your community that should remain secret.

Let the turtles have their sunny, sandy banks and the frogs and salamanders their clear and safe access to seasonal vernal pools.

Inappropriate Places and Vandalism

Sometimes the most interesting, out-of-the-way spots in the community have already been colonized by the local party crowd. No matter how interesting and benign they might seem during daylight, these spots should be avoided as Quest box locations. We once worked with a group of seventh grade students who quickly and enthusiastically zeroed in on a location for their Quest. We later learned that the reason these middle school kids thought the spot was so great was that all the older teens thought it was great too—for beer drinking and merrymaking. The treasure box was vandalized so often that we had to abandon the Quest. We had a similar experience with the Amphitheater Quest, which we ought to have predicted by the clues:

> Near the wall without graffiti there is a pipe on the ground,
> And within the deep darkness your treasure can be found.

Anne Swinscow, author of *Dartmoor Letterboxes,* told us, "Vandalism is not a great problem on Dartmoor. I suspect that your average vandal would rather vandalize cars in the car parks than walk out on the open moor! The only occasional vandalism is from the owner of a rival box who may very rarely take exception to a box for personal reasons . . . "

SELECTION PROCESS

Whether you make a rapid unilateral decision about site selection or involve your whole group in a thorough process of data gathering and prioritization will depend on your goals and your time frame. Where there is time, an in-depth selection process offers the benefit of educating your group about the community as a whole. It also offers an opportunity to strengthen your relationships with people from across the spectrum of the community, resulting in long-term benefits to your program. You can enhance the process by asking group members to visit some sites on their own before a final selection is made. You may even choose to visit and map several sites with your group before coming to your final selection.

If you have less time available, shorten the process by prequalifing some

4.4 Woodstock CWM treasure box. Photograph by Jon Gilbert Fox. What better place to end a Civil War Memorial (CWM) Quest than in the barrel of a cannon? The selection of a treasure box hiding spot is an important decision.

sites and presenting a shorter list to your group for a decision. In some cases, you may want to leave the site selection up to your community partner, if you have one. In addition to being a courtesy, this may result in your discovery of a new place.

HIDING PLACES

As you select the ending point for your Quest, keep an eye out for good hiding spots for the Quest box. In general, we have found that the best spots are hidden crevices or innocuous everyday objects that appear so common that they don't attract attention. In outdoor locations, we've had good luck hiding boxes in features we have modified or added such as:

➤ bird houses on trees or poles;
➤ bird feeders with false, pull-away bottoms;
➤ gray utility/maintenance boxes added to the sides of buildings;

- pots of plastic flowers;
- false mailboxes—as long as they are not really used for U.S. mail;
- compartments built into walls or fences;
- shelves built into the underside of boardwalks, seats, and stairs; and
- a sap bucket hung on a tree.

With indoor locations, we often involve the person behind the counter in some way. We've also successfully used:

- hollowed out books in a library;
- the menu holder in a restaurant;
- a brochure rack in a museum;
- the underside of a map dispenser; and
- a separate compartment inside a pooper-scooper dispenser.

Sometimes it's easiest just to hide the box in pre-existing structures or natural nooks. Just be sure that your spot doesn't involve any digging or disturbance, as this will quickly become a problem with repeated use. Here are some other hiding spots with which we've had good luck:

- beneath a heart-shaped stone in a store's garden;
- out in the wild, under a ledge, or inside a hollow log (bees also enjoy hollow logs, however!);
- in the crook of a tree;
- beneath an outdoor staircase;
- in a planter;
- on a crossbeam or roof support.

In places prone to vandalism, or just to add some fun, we occasionally place a padlock on the treasure box. It still needs to look uninteresting and innocuous—for example, a gray utility box. The padlock number or letter combination is gleaned through the clues. The clues might also reveal a password to say to the keeper of the key.

PERMISSIONS

We recommend that you obtain written permission from any public or private entity that is hosting a Quest box. Remember that this permission is best asked for and received *before* you begin the project rather than on the verge of its publication. While seeking permission may at first seem like a hassle, receiving that permission is crucial. On two separate occasions, months of student effort were wasted because Quests were made on private land without landowner permission. In many other instances, however, Quests became much richer—and the community was made stronger—because of the participation of a landowner.

It often works well to begin this relationship with a meeting rather than the formality of a letter. This gives you a chance to learn about and personally address any concerns before trying to finalize things. Be sure to discuss issues related to impact, parking, hours the Quest will be open to the public in different seasons, and publicity. Be clear about who the contact person will be for both parties. Following your meeting, summarize the discussion in a letter and invite their further comments on what you've written. Continue to correspond until you have reached an agreement that is satisfactory to both of you. If your Quest is in a park, talk to the department of parks and recreation. If you want to put the treasure box behind the library, ask the librarian. While you are at it, ask them for some stories about the place, and see whether they would be willing to check in on the treasure box every once in a while, too. Here is a sample permission letter:

Dear Friend,

I am the coordinator of the City Quest program.

The purpose of the City Quest program is to foster a *sense of place* in our region. To this end, City Quest is developing a series of treasure hunts to share our special places, landscape history, and cultural history with the community at large.

The City Quest program has two primary outcomes: education and stewardship. Quest-goers are *educated* about where they live and then, hopefully, *take care of* what they have come to know.

We are interested in developing a Quest at _____ that teaches about _____. We would like to hide our treasure box _____.

We hereby request your permission to:

➤ make a Quest at this location;
➤ publish the Quest in our City Quest booklet;
➤ hide a Quest "treasure box" at the aforementioned location; and

➤ have the Quest accessible to visitors during our Questing season (April 22–November 15).

All Quest-goers undertake these treasure hunts at their own risk. In granting this permission, we agree to hold you harmless regarding any liability stemming from accidents, actions, events, injuries, or litigation related to the treasure hunts.

Sincerely,

5) Varieties of Quests

THE CHALLENGE: GOOD PROCESS AND GOOD PRODUCT

The process of developing a Quest builds skills, sense of place, and relationships among various individuals and groups in the community. Along with offering a valuable experience for the Questmakers, though, the resulting Quests ought to be of high quality and have lasting educational and recreational value. The challenge is to find a way to engage your group in an authentic learning process that still results in an intriguing, attractive, and appropriately complex Quest. To that end, let's summarize the qualities that make a treasure hunt good . . . maybe even great.

It's fun. One thing is crystal clear: if a treasure hunt is going to compete successfully with television, computers, and all of the other myriad attractions and distractions of modern life, it needs to be fun, to surprise its participants and delight them, taking them joyfully away from their everyday concerns.

It's challenging. We all enjoy a good challenge, something that stretches our capabilities, tests us, and gives us a sense of accomplishment when we succeed. We tend not to be engaged by things that seem too easy. Great treasure hunts offer intellectual challenges: synthesis, deduction, and perhaps even some research. When appropriate, a little physical challenge such as a clue hidden on top of a hill or steep staircase can be a great addition, but only if there's a cautionary note about accessibility at the beginning.

It is ultimately solveable. Yes, we've just said that some challenge is good, but if all that work ends in frustration, the participant is unlikely to try again. Remember Helen Petersen's comment about the Pisgah Treasure? What if she and her husband had spent four months following flawed clues and had given up? It's unlikely they would have tried community treasure hunting again anytime soon. It's important that public treasure hunts be tested before their release.

It is aesthetically attractive. Lyrical clues evoking local heroes and events or half-remembered literary passages; whimsical illustrations hinting at mythical

5.1 Questing groundhog. Drawing by Patti Smith. Drawings, like this one taken from the *Valley Quest Map Book*, 1999 series, add a visually rich element of fun to your treasure hunts.

animals or half-hidden local sites—the best treasure hunts capture our fancy by appealing to our eyes and ears as much as to our sense of adventure.

It moves through interesting terrain. Whether leading participants through a city neighborhood where early civil rights marches were held or through a virgin hardwood forest to a boulder-strewn waterfall, good treasure hunts take participants to interesting and unexpected settings, often revealing the lesser-known facts and faces of the local scene.

It tells a good story. In the best treasure hunts, all of the elements—from the map to the clues to the treasure box—are linked together by a storyline. The story will be most effective if it has intrinsic value outside of the context of the hunt, hopefully living on in the memories of the participants and enriching their lives with local facts and associations. Good treasure hunts play off of the same theme again and again: in the title, in the text, in the map, and in the image on the hidden stamp.

It has engaging clues. Treasure hunt clue-writers can well take a cue from stand-up comedians, television commercial makers, and others who need to be able to grab public attention and communicate a small piece of information quickly. Staying tight to your story theme, using a little repetition, and embedding some local in-jokes in the text or map will get you a long way. While this may seem easier when you have a shared base of experience, as in *podchody*, it's also very possible once you get to know a place well. Calling attention to details also helps to inspire that "I've lived here twenty years and walked this street a thousand times and I've never seen that before" experience.

It reveals a human connection. Part of the appeal of the best treasure hunts is that some of the personality of the hunt's creators comes through. Delia claims that even though she has never met the creators of *Tracking the Dragon*, for example, she is sure that anyone who would offer a workshop on Water-

5.2 Secrets of the Forest Quest treasure box. Photograph by Steve Glazer. A cleverly hidden box is an important feature of a good treasure hunt. Here, a pulley mounted inside an old sugar maple lifts, and then lowers, the treasure box out of the cavity.

shed Cake Decorating, as they did, is going to be someone she would enjoy spending time with.

The treasure is cleverly hidden. While anyone who has thought about it knows that the main point of treasure hunting is the hunt itself, there's no denying that there's a thrill associated with actually finding the treasure box, too. That pleasure is even greater when the hiding spot is as clever as the clues were—

imagine finding an ornate, false-bottomed bird house, or a hollowed-out book on a library shelf, or a secret door in the side of a stone wall.

No two hunts are alike. The creation of one treasure hunt has a way of leading to the creation of others. One key to a successful program like the Dartmoor Letterboxes, is that there is *variety* in the designs, themes, and destinations of individual treasure hunts within the series. This helps maintain the interest of treasure seekers and builds loyalty toward the program.

We've found that one way to ensure that novice Questmakers include some of these elements of a successful treasure hunt—while still ensuring that they experience a rich and satisfying personal process—is to present them with a menu of successful models to use as a starting point toward developing their Quest. We call this the "Quest Taxonomy."[1]

QUEST TAXONOMY

When most people hear the term "treasure hunt," they think of an activity in which one clue leads to the next clue and so on. When we first started creating Quests with teachers, students, and community members, we found that almost every group assumed that we meant this basic format. In fact, in one of our early map books, we found that thirty-six out of thirty-nine Quests were of this type, which we now categorize as "Basic Clue to Clue." While some of these were, in fact, very good Quests, others were less effective and engaging.

We've seen an increase in both the depth and variety of Quests since we started presenting groups with a Quest Taxonomy. This taxonomy delineates a broad array of approaches to making Quests. While there are an infinite number of ways to create a treasure hunt, we have chosen techniques that seem to lend themselves best to Questing by meeting the following criteria:

➤ the Quest *design* can be accomplished by a group of children or adults working together;
➤ the experience of creating the Quest can have educational, social, civic, and/or recreational value for the people making it;
➤ the experience of going on the resulting Quest will have educational and recreational value for community members and local visitors; and
➤ the Quest is accessible by the public and can be used repeatedly without having to be set up each time it is used.

The Quest Taxonomy lays out a basic menu of options, and while you may choose one option from a single section, keep in mind the possibilities for adapting, combining, and adding to the different elements.

Quest Taxonomy

Section A: Quest Designs

A-1. Basic Clue to Clue
A-2. Elusive Clue to Clue
A-3. Letter Hunts
A-4. Number Hunts
A-5. Compass-Based Quests
A-6. Visual Detail Quests

Section B: Map-Clue Relationships

B-1. Full Map and Clues Set
B-2. Clue-Locator Map
B-3. Treasure Map
B-4. Mapless
B-5. Clueless
B-6. Clues Placed in the Landscape
B-7. Spurs, Floaters, and Other Variations

Section C: Quest Focus Areas

Place-Focused Quests

C-1. Meanders
C-2. Special Destinations
C-3. Special Routes
C-4. Special Means of Getting There

Story-Focused Quests

C-5. Cultural History
C-6. Natural History
C-7. Local Heroes and Heroines
C-8. Seasonal or Ephemeral Quests
C-9. Quests with a Mission

Quest Taxonomy Section A: Quest Designs

A-1. Basic Clue to Clue. This is the standard treasure hunt design, in which one clue leads to the next, and so on, to the treasure. The map sometimes has numbered locations that correspond with the numbers of the clue. This approach has the advantage of being a relatively simple design for beginner Questmakers to create—and for a beginner Quester to follow. It may also be a good choice when the focus is an involved story. Its major drawback is that often a clue-to-clue Quest can be solved by simply reading through the clues and the map—never leaving the comfort of one's armchair.

A-2. Elusive Clue to Clue. Clue to clue with a twist. This Quest design is similar to the Basic Clue to Clue except that it involves more problem solving. The relationship between the clues and the map may be less obvious, or the Quester may need to collect a number of the clues before the solution is clear, or solve a puzzle along the way. These techniques add spice and texture to the Quest experience while still keeping to a clue-to-clue format.

A-3. Letter Hunts. In letter hunts, the Quester gathers letters as he or she moves through the landscape. The letters may come from selected words on signs, designated objects, numbers translated into letters (1 = a, 2 = b, etc.), or even letters displayed by willing merchants along the route. The letters are assembled anagram-style to spell the treasure box location. Letter hunts work best in urban/built environments or in natural settings that have some interpretive or directional signage. If you design your Quest so that the letters don't need to be collected in a specific order, a large group can divide up and work on the Quest in small groups without tripping over each other.

A-4. Number Hunts. Our world is full of numbers and they make great Quest clues. You can find them posted as dates on buildings, addresses, route signs, even utility poles. Questers can also generate their own numbers by counting railings, doors, windowpanes, trees, steps in a staircase, or letters in a sign, or by measuring objects they encounter. Collected numbers can be used for everything from the combination on a padlock to the number of doorways to pass before finding the treasure box.

A-5. Compass-Based Quests. Some Quests are based partially or completely on the use of compass bearings. The level of compass use might be as simple as auxiliary reading to supplement what's already obvious from the map or as com-

PINNACLE HILL QUEST

29

To get there:

From the Lyme Green on Route 10, take Dorchester Road east. Go about 1.8 miles past the Alden Inn to the junction to Acorn Hill Road. Make a left onto Acorn Hill Road and go about 2 miles up through Lyme Hill and go about 100 feet further and look for the old dirt road on the left, section of Hardscrabble Road alongside the road. Where the Quest begins. *Please note that the driveways on the left at the intersection of Hardscrabble is private property—please do not use this driveway for parking.*

1 On this old dirt road, don't take a car.
   It's not safe. It's dangerous by far.

2 The trail is rocky and may be wet.
   I'd say hiking boots are the best bet.

3 Pass a yellow house on this trail.
   Keep going, and you will not fail.

4 At the fork in the road, don't go right.
   Just keep walking with all your might.

5 Unfinished road work you will pass by,
   Keeping straight is the route to try.

6 Open fields ahead and a pretty view
   Are just waiting to greet you.

7 Close to the woods you should keep
   Walking up but not too steep.

8 When you see two logs on a dirt pile
   Go right around them and walk up a rocky aisle.

9 Soon you'll pass more than one apple tree.
   Then you'll see an old chimney.

10 Facing the chimney's right front edge,
   Walk fifty-eight steps to the west.
   But take care not to fall off the ledge.
   The hiding place is . . . Have you guessed?

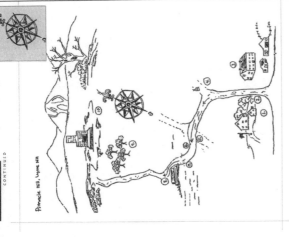

Pinnacle Hill, Lyme NH

Do you need some help or a clue?
Well, three steps is about four feet.
I hope that is helpful for you.
Soon you and our hidden box shall meet.

5.3 Pinnacle Hill Quest follows the basic clue-to-clue design. *Courtesy of Valley Quest: 89 Treasure Hunts in the Upper Valley.*

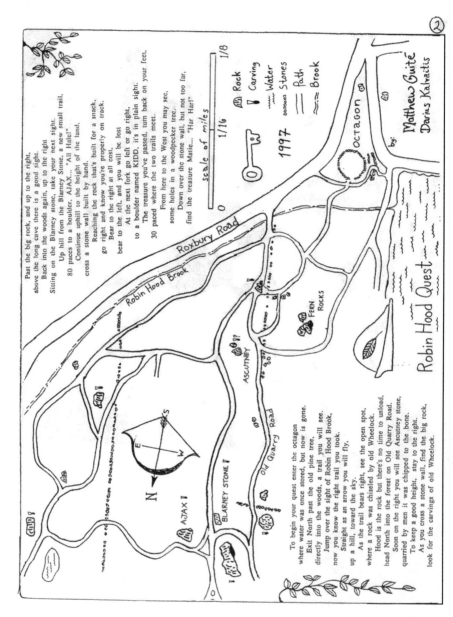

The map contains the following text:

Past the big rock, and up to the right,
above the long cave there is a good sight.
Back into the woods again, up to the right
Sitting on the Blarney stone, take your next sight.
Up hill from the Blarney Stone, a new small trail,
80 paces to a boulder. AJAX... "All Hale!"
Continue uphill to the height of the land,
cross a stone wall built by hand.
Reaching the rock that's built for a snack,
go right and know you're properly on track.
Bear to the right at all cost,
bear to the left, and you will be lost
At the next fork go left or go right,
to a boulder named KIDD, it's in plain sight.
The treasure you've passed, turn back on your feet.
30 paced where the two trails meet.
From here to the West you may see,
some holes in a woodpecker tree.
Down over the stone wall, but not too far,
find the treasure Matie... "Har Har!"

To begin your quest enter the octagon
where water was once stored, but now is gone.
Exit North past the old pine tree,
directly into the woods, a trail you will see.
Jump over the sight of Robin Hood Brook,
now you know the right trail you took.
Straight as an arrow you will fly,
up a hill, toward the sky.
As the trail bears right, see the open spot,
where a rock was chiseled by old Wheelock.
Hood is the rock but there's no time to unload,
head North into the forest on Old Quarry Road.
Soon on the right you will see Ascutney stone,
quarried by men it was chipped to the bone.
To keep a good height, stay to the right.
As you cross a stone wall, find the big rock,
look for the carvings of old Wheelock.

scale of miles
0   1/16   1/8

1997

Rock
Carving
Water
Stones
Path
Brook

OCTAGON

Roxbury Road

Robin Hood Brook

FEEN ROCKS

ASCUTNEY

AJAX

BLARNEY STONE

Old Quarry Road

Robin Hood Quest

by: Matthew Guité
Darius Kalvaitis

5.4 The Robin Hood Quest by Matthew Guité and Darius Kalvaitis is an example of an elusive clue-to-clue hunt.

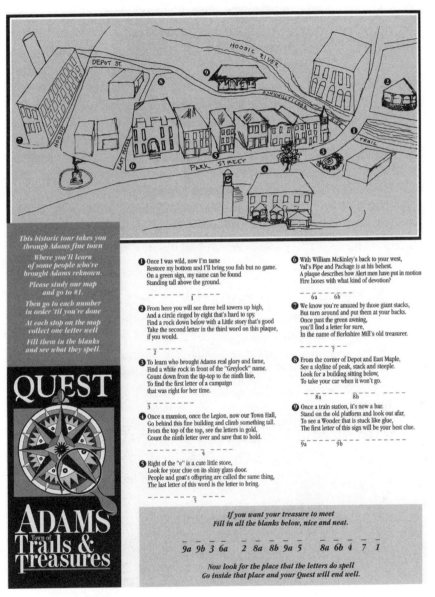

This historic tour takes you through Adams fine town
Where you'll learn of some people who're brought Adams reknown.
Please study our map and go to #1.
Then go to each number in order 'til you're done
At each stop on the map collect one letter well
Fill them in the blanks and see what they spell.

1 Once I was wild, now I'm tame
Restore my bottom and I'll bring you fish but no game.
On a green sign, my name can be found
Standing tall above the ground.

_ _ _ _ _ _ _ _ _ _ _
1

2 From here you will see three bell towers up high,
And a circle ringed by eight that's hard to spy.
Find a rock down below with a Little story that's good
Take the second letter in the third word on this plaque,
if you would.

_ _ _ _ _
2

3 To learn who brought Adams real glory and fame,
Find a white rock in front of the "Greylock" name.
Count down from the tip-top to the ninth line,
To find the first letter of a campaign
that was right for her time.

_ _ _ _ _ _
3

4 Once a mansion, once the Legion, now our Town Hall,
Go behind this fine building and climb something tall.
From the top of the top, see the letters in gold,
Count the ninth letter over and save that to hold.

_ _ _ _ _ _ _ _ _ _ _ _
4

5 Right of the "e" is a cute little store,
Look for your clue on its shiny glass door.
People and goat's offspring are called the same thing,
The last letter of this word is the letter to bring.

_ _ _ _ _ _ _ _ _ _ _
5

6 With William McKinley's back to your west,
Val's Pipe and Package is at his behest.
A plaque describes how Alert men have put in motion
Fire hoses with what kind of devotion?

_ _ _ _ _ _ _ _
6a   6b

7 We know you're amazed by those giant stacks,
But turn around and put them at your backs.
Once past the green awning,
you'll find a letter for sure,
In the name of Berkshire Mill's old treasurer.

_ _ _ _ _ _ _
7

8 From the corner of Depot and East Maple,
See a skyline of peak, stack and steeple.
Look for a building sitting below,
To take your car when it won't go.

_ _ _ _ _ _ _ _ _ _ _ _
8a   8b

9 Once a train station, it's now a bar.
Stand on the old platform and look out afar,
To see a Wonder that is stuck like glue,
The first letter of this sign will be your best clue.

_ _ _ _ _ _ _ _ _ _ _ _ _ _
9a   9b

If you want your treasure to meet
Fill in all the blanks below, nice and neat.

_ _ _ _ _ _ _ _ _ _ _ _ _ _ _
9a 9b 3 6a   2 8a 8b 9a 5   8a 6b 4 7 1

Now look for the place that the letters do spell
Go inside that place and your Quest will end well.

5.5 This Adams Trails and Treasures Quest is an example of a letter hunt.

5.6 David Millstone's Norwich Treasure Hunt is a visual detail Quest, searching out architectural elements. *Courtesy of David Millstone.*

plex as using compass bearings as your primary clues, combined with paces or other clues to indicate when you've reached the point where you take the next reading. The drawback of this approach is that it may limit the use of your Quest to those who already have a compass and compass skills.

A-6. Visual Detail Quests. Photographic clues or sketches are at the heart of this Quest design. The Quester searches within a prescribed area for a series of visual details. This design is a wonderful way to highlight architectural or sculptural styles and can work to teach about such natural history topics as tree identification in an interpretive setting like an arboretum.

Quest Taxonomy Section B: Map-Clue Relationships

The Quest-making approaches mentioned above can be effective in different situations, depending in part on how you combine the map and clue elements. Various combinations lend themselves best to different teaching or community-building goals, to different Quest focus areas, and to the different local conditions you encounter. The most-utilized map and clue relationships include the following.

B-1. Full Map and Clues Set. This is your basic belt-plus-suspenders operation, in which a full Quest-area map duplicates or reinforces a full set of matching clues. It is the most common kind of Quest today. With this combination, Questers can generally get where they are going by following the clues alone, and the map is either for decoration, to point out sites of interest, or for general reassurance.

B-2. Clue-Locator Map. The map and the clues each have a separate function in the clue-locator map model. The Quester's movement is entirely directed via a full Quest-area map that identifies each stop. The Quester follows the map to a series of designated stops, then reads the clue that is assigned to that location. The clue may tell a story about that particular spot or may direct the Quester to collect a number or letter to use in finding the final location.

B-3. Treasure Map. In this model, clues direct the Quester's movement to each stop. The map is a very local map pinpointing the area immediately surrounding the treasure box. This map kicks in only after the Quester has reached the final treasure location. The treasure map may be self-explanatory once the Quester has reached the site, or it may be used in conjunction with letter or number clues, a final hint clue, or a visual clue.

B-4. Mapless. In some cases it makes sense to eliminate the map altogether, in favor of more space for a lengthy saga and clues, or when there is a focus on visual details such as photographs, sketches, or historical documents. In these cases, the clues guide the movement alone or sometimes in conjunction with the graphic material. A variation on this is the mobile box, in which clues lead the Quester to a certain ice cream truck driver—or other genial community member—who agrees to carry the treasure box in his or her truck, relinquishing it for sign-ins only upon a password.

B-5. Clueless. Clueless Quests may feature only a detailed map, though it may have written clues artfully embedded in it. Once the box is found, there is often a suggested activity, question, or challenge associated with the site.

B-6. Clues Placed in the Landscape. Rather than relying on the situation as it first presents itself, some Questmakers alter things slightly to suit their purposes. This might include printing letters or numbers and displaying them in shop windows or painting them on blocks of wood, attached to a cord and hidden out of sight in the crook of a tree or behind a boulder on a trail. It might

SHAKER VILLAGE CEREMONIAL SITE QUEST 10

To get there:

Take I-89 south of Lebanon to Exit 17. Take Route 4 east to Route 4A. Make a right onto Route 4A. Drive south 3.3 miles. Park in the driveway with a sign for Stone Mill on the right (west) after passing the Shaker Museum.

1 Look west toward the hill and you'll see a sign that shows Enfield's Conservation line.

2 Follow the well-beaten path up the hill.

3 Right at the fork to stay on it still.

4 Where the trail turns left, ruins of a bridge you'll see to a field where merino sheep used to be. (Pass the bridge, staying on the trail).

5 The trail turns northwest, as you go, with a beautiful view of the lake below. (Observe bluebird houses).

6 Across the top of the hill and then you'll see an old granite post and a big locust tree.

7 Walking on, look down. Below was once found the end of a run where skiing was done.

8 At the little white post, take a short break for a compass reading you may take. Head west at the bearing of 268 (Hint: due west is 270 degrees).

9 Go a hundred yards heading west.

10 You're almost there—the end of your Quest. March up the hill like a Shaker would and come to a grove set back in the woods. You'll see a gate and a sign.

11 Here is a place of quiet and prayer Take some time of reflection whilst there. At the southeast corner of this spiritual place you'll find our box's special space.

Hint: Granite Post marks SE. Corner

KEY
Treasure Chest
Bluebird House
Bridge
Granite Post
White Post

5.7 The Shaker Ceremonial Site Quest features a full map and clues set. *Courtesy of Valley Quest: 89 Treasure Hunts in the Upper Valley.*

79

THE GEORGE, FREDERICK, AND LAURANCE QUEST

To get there:
This Quest begins at the village green in Woodstock.

Please study our map and go to #1.
Then go to each number until you are done.
At each stop collect one or two letters as well,
Fill in the blanks and then see what they spell.

1   Our Quest starts on the south side of the green
Where a historical marker sign can be seen.
Woodstock was the site of the very first one
Find this word on the sign and get set for fun!
First letter of this word: __ __ __ __ __ __
1

2   Now ladies and gentlemen, girls and boys,
Go inside this fine building without any noise
Beside the fireplace on the wall there's a rug
Find your letter then give your friends a big hug.
First letter of the second word on the rug: __ __ __
2

3   Down Bond, last on left, Billing's #4,
Look for the tiny sign beside the door.
Second letter of first word on the sign: __ __ __
3

4   Frederick Gillingham ran this store,
On the big sign find out who helped him with this chore.
Second letter of the last word on the sign: __ __ __ __
4

5   The Rockefellers were married here many years ago
To the main sign by the driveway you should go.
First letter after the word "Vermont": __ __ __ __
5a
Second letter of the second word on the sign: __ __ __ __ __
5b

6   Go to the left of the building to the Paul Revere Bell,
Look at the sign on the bell stand and it will tell.
Your next clues are inside this visitor center,
Follow the signs and then you should enter.
Go to the sign that's straight ahead with great speed

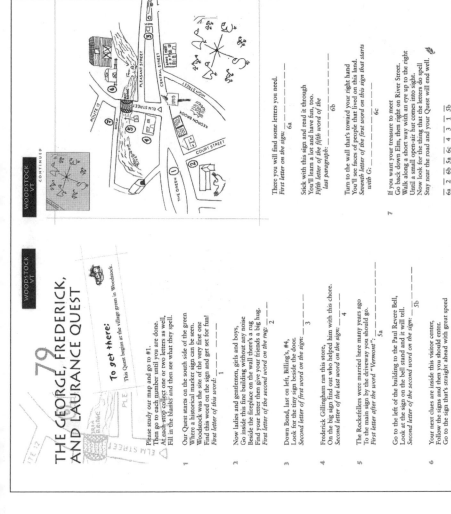

There you will find some letters you need.
First letter on the sign: __ __ __ __ __
6a

Stick with this sign and read it through
You'll learn a lot and have fun, too.
*Fifth letter of the fifth word of the
last paragraph:* __ __ __ __ __ __ __
6b

Turn to the wall that's toward your right hand
You'll see faces of people that lived on this land.
*Seventh letter of the first word on this sign that starts
with G:* __ __ __ __ __ __ __
6c

7   If you want your treasure to meet
Go back down Elm, then right on River Street.
Walk along a short way with an eye up to the right
Until a small open-air hut comes into sight.
Now look for the thing that the letters do spell
Stay near the road and your Quest will end well.

__ __   __   __ __   __ __   __ __   __   __   __   __ __
6a   2   6b   5a   6c   4   3   1   5b

5.8 The George, Frederick, and Laurance Quest is an example of a clue-locator map relationship. *Courtesy of Valley Quest: 89 Treasure Hunts in the Upper Valley.*

88 VILLAGE GREEN QUEST E

Start here

To get there:

Take Exit 1 from I-89 and follow Route 4 west for about 10 miles to the village of Woodstock and the green. This Quest begins at the Woodstock green.

Stand at the Info Booth on the green
Where two nearby mountains can be seen.
With Mt. Peg on your left—small, green, and round,
And Mt. Tom on your right—tall, cliffy, and brown,
Start on down the path, headed straight out of town.

At the very first crosswalk to the right that you see
cross over and walk left to solve our mystery.

Stay on the sidewalk to the right of Route 4,
go to a place where you can see movies galore.
This is also a place for all our town meetings,
Where neighbors and friends can exchange friendly greetings.
Number of big columns on the front of this building = ☐
                                                          A

Stay on Route 4 as it curves out of sight,
'Til you come to a building that's totally white.
Don't worry, don't cry, you'll know that you're there,
When you see this white building point high in the air.
Number of matching doors on the front of this building = ☐
                                                            B

Continue on to a place where you can stand on air
And look over the edge if you want a big scare.
Listen upstream and to the right for a hint of your glory.
Number of thin, round poles in each section of the railing = ☐
                                                                C

As Route 4 curves to the left, stay by its path,
Cross "River" then "Mountain," then prepare for some math.
At the crosswalk up to your left, cross over Route 4,
You're getting very warm, just a little bit more.

Map for Final Clue

Key

'Look over the edge if
you want a big scare.

Listen upstream and
to the right **for a hint**
of your
                g l o r y'

Take your answer for "A" and add it to "B,"
Together they'll give you an answer that's "D."
Now subtract "D" from "C," but don't tell a soul,
Go to this number on your map, you'll be at your goal!

☐ + ☐ = ☐
A   B   D

Then

☐ − ☐ = Magic Number!
C   D

We hope you've enjoyed our Valley Quest,
If you keep it a secret, we'll think you're the best!

5.9 The Woodstock Village Green Quest E features a treasure map. The clues guide you to a treasure map that reveals the location of the treasure. *Courtesy of Valley Quest: 89 Treasure Hunts in the Upper Valley.*

also include hiding a box partway through the Quest that contains clues for finding the final Quest box.

B-7. Spurs, Floaters, and Other Variations. Once a Questing program gets established in your community, you'll find that people start coming up with all kinds of ingenious variations, sometimes implementing them without your prior knowledge. How about a "Spur Quest," for example? Someone explores the area around where your box is hidden and finds a nearby marvel—an added bonus! They create a map and clues to this additional feature that are available only in the treasure box of the original Quest, which must be solved first. Another Quest mutant is the creation of a "Floater Box." Like the cowbird, which lays its eggs in other birds' nests, these parasites are placed randomly in or next to other Quest's treasure boxes. They include directions to those who come across them to sign in and then place the box in the treasure box of the *next* Quest they solve, wherever it may be. "Rovers" are boxes carried by Questers while Questing; they are only discovered through chance meetings in the field. Far from being disrespectful, we find these variations to be a good indicator that the public is making the Questing program, and the landscape, their own.

Quest Taxonomy Section C: Quest Focus Areas

Most Quests fall into two categories of focus: those inspired by a particular place and those inspired by a particular story or message. These often happily entwine with each other. Here's a list to get you thinking:

Place-Focused Quests. Quests are often inspired by a sense of the special places in the community.

C-1. Meanders. Many Quests, especially those created by beginning Questmakers, are focused merely on taking people out and about and giving them a fun walking experience and a nice introduction to the neighborhood. These meanders often include a historical fact or two, maybe a cultural reference or a nature note, a visit to a local point of interest, or a stroll down a trail, all linked together by being near enough to each other to be in the same Quest.

C-2. Special Destinations. If the first thing you hit upon when conceptualizing your Quest with your group is "Let's make it lead to McGonagle Tower!" and that idea just doesn't go away, then you're probably going to be looking at a special destination Quest. The story, the route, and the hiding spot all contribute to the celebration of this one amazing place.

C-3. Special Routes. In some Quests, the focus is less on the destination and more on what you see along the way—sort of a grand tour. These special routes differ from meanders in that they're linked by a thematic focus—for example, looking at a certain type of architecture, the locations of past libraries, or a string of examples of damage caused by the great flood.

C-4. Special Means of Getting There. A particular means of transportation is sometimes the initial spark for these Quests. They range from Quests meant to be done by bike to Quests that can only be completed by boat or on skis. Other Quests may take advantage of a community's public transportation system or may be on mountain paths that can only be hiked. These Quests may be their own kind of meander or may be developed to portray a particular theme.

Story-Focused Quests. An interesting local fact or event can serve as the integrating theme for a Quest:

C-5. Cultural History. Well-known historical events of national importance, little-known happenings peculiar to your community, and everything in between make great fodder for Quests. Even a short Quest can get into enough detail to spin a good yarn, whether it explores the story of a mysterious old foundation, the time when gas lamps lit your city, or the impact of the Lewis and Clark and Sacagawea expedition on your town. Stories about local heritage often work especially well for historical groups or teachers doing social studies units.

C-6. Natural History. Quests are a great way to teach local ecology, biology, geology, and other features of the natural landscape. Natural history Quests might ask: Why is the water in this creek brown? What makes rocks crack or move? How do you tell a white pine from a red pine? When do the spring wildflowers bloom in this region? The answers come right out of the textbooks and into a vibrant explorer's game. These Quests offer a great opportunity to partner with your local nature center or museum.

C-7. Local Heroes and Heroines. Local people who've done great things, not-so-great things, or even bad things, or who simply lived their lives independently enough to grow into a small legend—their stories can make fascinating Quests. The routes often lead by their former homes or the scenes of their greatest accomplishments or misdeeds. The clues hint at their stories, which can be embellished in more detail by a sheet in the Quest box, hidden at a point of great significance to their lives' stories.

67

LINNY'S LOOP BICYCLE QUEST

To get there:

Travel on Vermont Route 113 to the village of Post Mills. Then take Route 244 north along the west shore of Lake Fairlee (this is Lake Fairlee Mills). At the foot of the Lake Fairlee access/boat camp 1.8 miles from the Lake). Your Quest begins there!

Welcome to our lake! Fifty-six feet at its deepest point—and six miles around.
Looking carefully, its seven in-flowing sources can be found.
Today we'll explore just two—leaving the other five up to you.
Hop on your bike and head northwest (left) to look
At the tributary known as Middle Brook.
At the bridge, stop and take in the view
Of all the nutrients flowing under you.
Nutrients, perhaps, that you cannot quite see . . .
But what about birds, beaver or otter? What do YOU see?
Eagles? Osprey? Muskrat lodges? Wood ducks? Painted or snapping turtles?

From the bridge find the white arrow. Take the first word
And fill it in the space labeled #1. That is how this Quest is done!

After a right turn on Middle Brook, to your left is a camp
And along the banks of the brook a willow alder swamp.
Now take a ride in style . . . for at least another mile.

Next to 1003 is something with many, many windows.
Fill that word in space #2, for that word is your next clue!
One side of this fine structure is marked clearly with two Xs.
The number of windows on this side goes in space #3. Really!

You'll pass a lake on your left and then the three silos of Stever's Dairy,
Still milking and shipping milk—which makes our hearts merry.
There is a breed of cow named upon their sweet sign.
Fill it in space #4 and you're doing just fine.

Three miles in, watch out for the griffin!
Upon a lawn on the right you may see him a-sittin.
One part eagle, one part lion; lots of presence but hardly flying,
Straight on lies a church marking West Fairlee Center
Bear Notch Road turns left but *do not enter.*
Stay on the main road right and straight—
For Blood Brook Road is the name of your fate.
Turn up to the right when this road comes in sight.
The climb may be steep, but try and stay on your seat . . .
It's only a half-mile climb. Breathe, relax, take your time.
The reward comes as a long downhill without a clue . . .
For a little while the entertainment's all up to you!

There's a peaceful resting place under some trees,
And the top of its fence is marked clearly with these.
Mark this symbol as word #5—then off you drive!

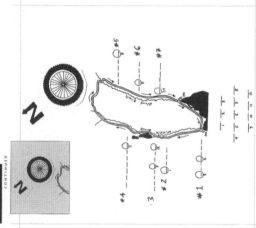

You'll pass a one-room schoolhouse from 1871
Record the name of its number and clue #6 is done.

Take a right on Marsh Hill to see a near old sawmill.
Lichen has covered the rocks in its cellar
Turning them to a quite curious color.
The seventh word names the color you've learned.

Back to the main road and coast down to the lake
And at the stop sign a right turn you must take.
Here, Blood Brook flows in through a culvert
Offering the lake rich freight for dessert:
Rock flour, pulverized leaves, soil, and even pieces of trees.

Alas, the time has come to compute and to spell . . .
And if you've collected all the right words you can tell
The place the Linny Loop Treasure Box will dwell.

Missing two clues? Aye, use your eyes!
Then bike .3 miles until you see
The triangle where the treasure must be!

_____ _____ _____
4   5   6   7   8

_____ _____ _____
9 10 11 12

5.10 Linny's Loop Bicycle Quest makes a nine-mile circuit through the Middlebrook and Blood Brook valleys. *Courtesy of Valley Quest: 89 Treasure Hunts in the Upper Valley.*

HARTFORD RECYCLING QUEST 25

To get there:

Take I-91 to Exit 11 in White River Junction. Go north on Route 5 for about 2 miles. When you pass the Maple Leaf Motel on your left, watch for the Recycling and Hartford Community Center just ahead on the left (east) side of Route 5. This Quest can be done 8 to 4 Monday through Saturday.

Your Quest starts at the parking lot,
Walk to the display where many things rot.
Full of rotting leaves, apple cores, and weeds,
Compost provides what a garden needs.
Don't throw your food scraps in the trash!
Compost is like gold—as good as cash.

Facing the bins, turn to the left,
Enter a room and follow your Quest.
Look at the walls of this entryway—
It was children who made this display.
What did they use to make the forest creatures?
The eyes, the noses—and the other features?

Open a door to see a dragon with wings:
She's made of plastic jugs and other things.
Get on the bike and pedal 'til you're sore.
Which takes more energy? Recycling or mining ore?
There're lots to do and learn in this room,
Have fun now—for your next stop is a tomb!

Follow the porch and cross the drive here—
But look both ways to make sure that it's clear.
Look at the land going up a slope,
There's a stone here that's kind of a joke.
It is a tombstone for a mountain of trash,
A huge, mummified garbage stash.

Now turn around and read the signs.
We collect hazardous wastes here—all kinds.
Bleach, batteries, chemicals to kill aphids and ticks—
Don't throw 'em in the trash! They'll poison the groundwater,
and make us sick.
Now stop, close your eyes, and please think . . .
What hazardous waste is under your sink?

Turn back and carefully cross the drive again.
Follow any people you see carrying bins
To a room where we take used cans, bottles, and paper.
It's all sent away to factories where they remanufacture—
New cans, glass, and paper for all to use.
Recycling saves resources and reduces refuse.

Want to see the cans and bottles fall down?
You're near the treasure so don't clown around.
Enter the far door to the "Good Buy Store."
Here are used clothes and trinkets galore!
Look for the "observation balcony" sign,
Enter, and look around for what you can find.

The treasure box is very near.
It is in the corner—so your Quest ends here.
Hope that this Valley Quest was a blast
And you learned to choose disposal last.
It was a pleasure to have you here as a guest.
Remember: "Reduce, Reuse, and Recycle" are the best!

5.11 The Hartford Recycling Quest is a Quest with a mission. This Quest introduces many people to the town of Hartford, Vermont's educational recycling center. *Courtesy of Valley Quest: 89 Treasure Hunts in the Upper Valley.*

C-8. Seasonal or Ephemeral Quests. From blooming orchids, spawning suckers, and fruiting cactuses to migrating salamanders, swirling skaters, the Annual Strolling of the Heifers, and the Rutabega Curling Festival, there are myriad things in each community that happen only once a year, and for a short time at that. Seasonal Quests highlight and celebrate these ephemeral oddities, enhancing their educational and place-making value. We have found that impressions from the stamps of these ephemeral Quests have become highly coveted items, building anticipation for the arrival of the box the next year.

C-9. Quests with a Mission. These Quests are designed with a mission in mind—to get Questers to recycle, clean up litter, donate their used clothes, or walk more and drive less. Often created in conjunction with a local environmental or social organization, these Quests offer a fine opportunity for developing community partnerships. Beware, though—it can be tricky to strike the right balance between authentic community service and over-the-top advocacy. Test your idea with diverse people in your community to be sure your message is on target.

Depending on your group's goals, you may want to combine any number of these options into a diverse offering of community Quests, such as the South Shore Quests. Or you could create a series of Quests that hold together tightly in both theme or structure—Environmental Heroes and Heroines of Lowe's Creek, for example, or Biking Quests of the Washington River Corridor. We suggest that you start by opening your idea bank wide; then, once you have as diverse, even weird, a set of ideas as possible, hone, tweak, and combine until you have the best of the best: what is just right for your place.

6) Notes for Teachers

CREATING A QUEST CAN BE an engaging activity for a classroom or school group, serving either as an integrating project for interdisciplinary studies or a focused project within a particular discipline. Students ranging from second graders through college age have had solid learning experiences while making successful Quests. No matter the age of your group, however, the foundation of a Quest consists of the following educational practices.

1. Adopting a particular place in your community.
2. Visiting that place again and again.
3. Out in the field, learning to observe details and discovering the characters and stories that inhabit that special place.
4. Back in the classroom, using these details as a basis for integrating diverse academic disciplines: art, writing, reading, mathematics, science, social studies, and technology.
5. Utilizing community resources and partners like the historical society, town office, and library; deepening the Quest participant's understanding through primary and secondary source investigations.
6. Inviting the elders and specialists in the community to participate in your Quest. Making use of the content mastery that already exists in the community. Building relationships across the generations.
7. Publishing and distributing your Quest in order to share the student's learning with the broader community.

PREPARATORY STEPS

Before creating a Quest with a group of students, go on a Quest yourself or take a Virtual Quest on the Vital Communities or Antioch New England websites. Doing this will help you get a feel for a Quest project's final outcome or product. Then, review your curriculum to determine where and how your particular curricular goals might align with a specific site in your community.

Forming a Team

Next, meet with your colleagues to see if a Quest is something that they might wish to undertake as a team. Create a curriculum map, brainstorming all of

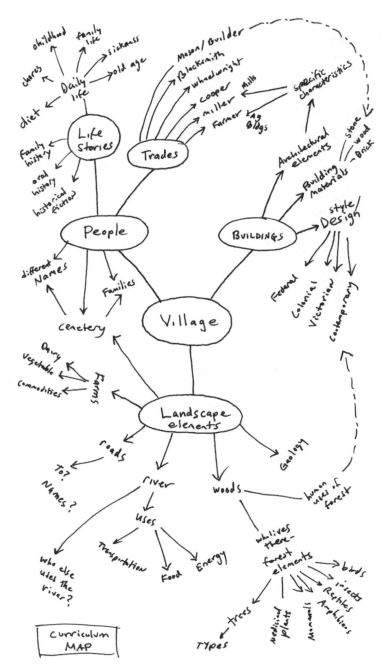

6.1 Use a curriculum map to brainstorm your ideas and resources, and to align goals and activities into a coherent and effective project.

your curricular needs, site options, and goals. Review all of those ideas, highlighting and connecting the most important components. There are many possibilities, but focus is good.

Would the gym teacher like to work with the students on orienteering, so they can learn to take compass bearings? Would the art teacher enjoy teaching a sequence that starts with journal making, moves through field drawing and found object sculpture, and ends with lino-cutting stamps of the trees of your forest? Does your school have a technology coordinator? There is plenty of room for technology in creating a Quest: databases, digital cameras, maps with GIS/GPS technology, and desktop publishing. Align as many interests as you can.

If you are an elementary or middle school teacher, you may choose to focus your Quest within science or social studies, but try your best to approach the project as a curricular unit incorporating multiple disciplines. The Quest process works best when each discipline informs and reinforces the others: when you are writing about the animals you are observing but also drawing them, measuring them, learning their diets, and collecting their scat, too.

Time Commitment

Once you have decided to take on a Questing unit, you'll want to have three or four weeks for preparation time. This period should include at least one or two visits to your chosen Quest site on your own (or as a team) to preassess and address any challenges or concerns. You'll also want to line up—well in advance—your volunteers and fieldwork dates.

Six to eight weeks are generally required to complete a Quest unit if you work on one or two lessons a week. Some schools choose to work on Quests more intensively. The Woodstock Elementary School has had good success making Quests every year over the course of the last two weeks of the school year, after students have finished their Vermont History unit and before they leave for summer vacation. This school has made more than a dozen Quests in seven years. Most of the Woodstock Quests begin at the Woodstock Village Green, less than a hundred yards from the school. Your destination need not be far away.

Field Trips

In general, Quests will require field trips, for your group's initial site visit, for drawing of map elements, and for drafting movement clues. An additional field trip or two will be required for research, including perhaps a visit to the historical society or an oral history gathering. On your final visit you can test the Quest and plant the treasure box. During some weeks (especially in the second half of the project) an additional period or two of in-class work time might be required. Budget a half day for each field trip.

Your fieldwork may happen on or adjacent to your school campus. If you are traveling further afield, you will have to depend on cars or buses. Note that many successful Quests made by students begin within walking distance of schools. If you are going to be traveling off campus, remember to send out permission slips well in advance. We often send out a "blanket" permission slip covering the entire project. Going further afield may require volunteer drivers as well.

Volunteers

In general, the more volunteers you have for your field trips, the stronger and deeper your Quest will be. In the best of all possible worlds, you should try to have one adult for every four to six students. Know that it is well worth the extra effort to attract at least one or two volunteers from your community. To have one parent and one community member participate will yield rich results.

Build a Plan

Hold a follow-up meeting to continue to brainstorm your site, content, and goals, working towards nesting individual lessons in a calendar or timeline, for example:

Lesson #	Activity	Date
1.	Reading *Your Own Best Secret Place* and brainstorming our special places	4/25
2.	Making journals	4/30
3.	Site visit # 1: general exploration	5/2
4.	Site visit # 2 with the art teacher	5/9
5.	GPS lesson in class	5/15
6.	Site visit # 3 GPS with technology coordinator	5/16

and so forth.

Be sure to distribute the calendar to all participants.

Unit Rubrics

Based on your curriculum and planning document, we recommend that you develop two unit rubrics—one unit for your team, the teachers, and a second, simplified unit to distribute to students. It is important that students have an understanding of the entire project, and see how all of the different activities and pieces fit together. See the appendix for a sample of a teacher's unit rubric. Note that each lesson is linked to a particular standard and culminates with a product or performance. Each product or performance, in turn, references an evaluation tool. Students can be given something more like figure 6.2.

As you think about goals and outcomes for the project, also think about the content areas your students will be delving into. Many teachers find it helpful to create pre- and post-assessment questions as a way of clarifying and quantifying their student's learning.

Sample pre-assessment questions focusing on elementary school geography:

1. What river forms the western boundary of Cornish, N.H.?
2. What state borders Cornish?
3. Name three hills in Cornish.
4. Place them (approximately) on a map of Cornish.
5. Name three brooks in Cornish.
6. Place them on the map of Cornish above.
7. What is the town that borders Cornish to the north?
8. What is the town that borders Cornish to the south?
9. What is the town that borders Cornish to the east?
10. What is the town that borders Cornish to the west?

A sample pre-assessment focusing on forestry might include the following questions:

1. How many coniferous (evergreen) trees can you name? What are some of these trees' key distinguishing features?
2. Name three trees of the "understory." Give some of their distinguishing features.
3. Sometimes, if we look closely, we can see "signs." Give some signs of the following:

   Insects

   Weather Incidents

   Animals

   Humans

Project Overview: **We will be designing a "Quest" treasure hunt that wanders through our cemetery and teaches visitors about our town's history.**

Our activities will include:

_____ Creating a GIS treasure hunt map
            Each student will contribute 2 GPS "points"

_____ Creating rubbings of gravestones
            1 complete rubbing per student

_____ Digital photographs recording stones in the cemetery
            2 photographs per student

_____ Clues, written in verse, which teach other what we learn
            1 4-line "teaching clue" written by each group
            1 4-line "movement clue" written by each group

_____ A data record, cataloguing information for each stone in the cemetery
            2 complete data records per student

_____ A database, containing all of the data records listed above
            2 data records entered by each student

_____ A complete treasure box, including:
         ____ a sign-in guest book
         ____ a hand-made stamp
         ____ an ink pad & pen
         ____ laminated copies of old maps, photographs, stories, etc.

Quality standards/ rubric:

Activity	Below Standard	Meets Standard	Exceeds Standard
Rubbings	incomplete or sloppy	complete	complete, carefully done
Photos	0 or 1 photos taken	2 photos taken	taken and accurately labeled
Movement Clues	incomplete or poorly composed w/ inaccurate directions	complete with accurate directions	thoughtful, well composed, accurate directions
Teaching Clues	incomplete or poorly composed insufficient information	complete with relevant information	thoughtful, well composed, rich with information
Data Records	Incomplete records	Complete records	complete, accurate and carefully written
Database	Task not completed or data is not accurate	Task complete with accurate data	Task complete, accurate assists with graphing data

6.2 Provide students with an overview of your Quest project including its goals, student-created products, and your criteria for evaluation.

You can undertake the pre-assessment formally or informally, individually or in a group setting. If you repeat the assessment after your Quest, you'll be able to see evidence of content knowledge.

GETTING STARTED WITH YOUR CLASS

We have had success easing into a Questing unit with student groups using three distinct approaches:

- ➤ telling or reading stories about treasure hunts or special places;
- ➤ studying community maps;
- ➤ brainstorming on an oversized sheet of paper about the most important features of the community.

Opening Stories

Storytelling is a wonderful way to begin a project. We have used a variety of different books—special place books, treasure hunt books, regional nonfiction, and historical fiction.

Special place books evoke the power of specific environments. Our favorite special place book is *Your Own Best Secret Place,* a picturebook story of discovery by Byrd Baylor (illustrated by Peter Parnall). One child finds the abandoned special place of another—a hollow oak tree. Inside the tree, the reader discovers a can filled with a pencil, notes, drawings, a blanket, and a candle. Love and stewardship are passed on from one generation to the next in this heartening story. You can read the book out in the field or in the classroom. Either way *Your Own Best Secret Place* raises all of the key ingredients of a Quest: discovery, investigation, stewardship, the arts, and love.

There are a lot of good treasure hunt books as well. *The Seven Treasure Hunts* by Betsy Byars is a story about two friends who take turns making treasure hunts for each other. With each chapter the treasure hunts become more complex, as the children learn about what works and what doesn't. *The Seven Treasure Hunts* is good fun—with a lot of practical informationas as well.

We also love *The Treasure* by Uri Schuleweicz. In this folktale, the protagonist has a recurring dream that gives him clear instructions on where to go in order to find a priceless treasure. The man shrugs off the dream twice, but after his third dream he heads out on his Quest. You can easily adapt this story to reflect the special qualities of your place by changing the protagonist's name, his destination, and the landmarks. The message—so beautiful—is universal:

the treasure, everything we need, is right here before us, if only we know how to see it.

Look for historical fiction and nonfiction books that focus on the story you want to explore through your Quest. *The Ox-Cart Man* by Donald Hall, for example, teaches about village life before cars and helps get students into the mindset of a time and world unknown to them. By beginning with this kind of story, your group can start to conceptualize the narrative structure that will later unfold, in clues, over the course of a location in your community.

Charting Key Features of Your Community

Another way to begin your unit is to brainstorm the elements of a map of the special places in your community. Start with an oversized blank sheet of paper or an empty chalkboard. Add the cardinal directions, then follow the directions for Community Mapping in chapter 4. Charting key features can also serve as a good pre-assessment, allowing you a glimpse of how much or little your students know about the geography and cultural life of their community. If you ask your group to do this again at the end of the project, you can see how far they have come regarding their knowledge of their home territory.

For more details on using community maps as an entry point, see the first lesson in the appendix, the Village Quest Unit.

Making Journals

Early on in the project, it is a good idea to have students make field journals in which they can record all of their work. Students can keep day-to-day journals of their activities; they can use them out in the field for sketches and field notes; they can record notes, ideas, and questions as they undertake research; and they can use them as workbooks for drafting riddles and clues, and sketching map elements.

By undertaking field journal exercises on every site visit—and when wrapping up each activity or meeting—students begin to understand the role of writing in learning and the place of journal keeping as a part of lifelong and place-based exploration. Remember to share work from your own journal, as well as inspiration from the journals of others: Gilbert White's Selbourne journals, and Hannah Hinchman's *A Trail through Leaves* are two fine examples.

You can have your students make their own journals by following the instructions given for making sign-in books in chapter 11. Better still, have your art teacher or a local artist partner join your group to make artistic journals reflecting the themes with which you are working.

6.3 Field journal created by fourth graders at the Lyme School. Making and keeping field journals enriches the artistic and literary aspects of the Quest.

TAKING STUDENTS ON A SAMPLE QUEST

By creating a very simple sample Quest, you can give students the experience of what a completed Quest feels like—and make clear to them the core components and outcomes: the clues, the map, the treasure box, and the box components. You'll need:

A map of your school
A treasure box (any kind of box for an inside treasure, perhaps a plastic
    sandwich box for an outside treasure)
A small spiral notebook
A pen
An ink pad
A simple rubber stamp
A magic marker
A dozen or so index cards
An equal number of envelopes
Scotch or masking tape
A copy of the clues listed below

Sample Quest Clues

When people come to visit your school,   (main office)
They come here first—that's the rule.

Line up here and do not roam,   (bus pick-up)
Big yellow things will take you home.

No food, no noise, but lots of books.   (library book drop)
Where you return your stuff—that's
    where to look.

Keeping time with scoreboards or clocks,   (school gym)
Players with sweaty bodies and smelly socks!

Go find the one who rules your school   (principal)
Who you'll get sent to if you act the fool.
But today just ask for something cool.

The one who cleans the windows and floors,   (supply closet)
Keeps the supplies behind these closed doors.

Bumps and bruises, sick or cut,   (nurse's office)
Come in here to get fixed up.

Your school's name is written here,   (school nameplate or sign)
For all to see from far and near.

Outdoors there's a place to swing   (swings)
Where you can have fun until the bell rings.

Waving proudly in the breeze   (flagpole)
Salute my colors if you please.

A stream of liquid to quench your thirst   (water fountain)
After running around you come here first.[1]

Steps for Making the Sample Quest

Choose a hiding place from the list above—for example, the library book drop.
Place the treasure box there, filled with a greeting, a pen, the sign-in book, the
stamp, and the ink pad. Now go to your next location—for example, the flagpole. When you get to the flagpole, put the last clue—the clue that leads to
the book drop—in an envelope and tape it to the flagpole. Note that you are
working your way *backward* through the treasure hunt. At each new location,
you are leaving the directions to your previous location. When you have placed
as many clues as you wish, go back to your classroom or starting point. You can
now begin your Quest with the riddle that leads to the last clue you hid—the
first of their Quest. This clue can be hidden too, if you create your own riddle
clue that points toward its discovery.

Placing the clues in the envelopes helps slow down the process, allowing
you to continually regroup and spread the success around to a greater number
of participants. Some individuals can have the thrill of discovery, while others
can take turns reading the next set of clues aloud.

Be sure to have a treasure box at the end. When students discover this and
open it up, you will have a perfect opportunity to point out the roles played by
the stamp, stamp pad, and sign-in book.

Penny Treasure Hunts

Students like hiding a treasure as much as they do finding it, so this Penny Treasure Hunt is a great way to inspire their interest and develop their mapping skills.

Give each student or pair of students a penny. Make sure each penny has a different date on it. Hand out clipboards and white unlined paper, and make sure everyone has a pencil. Ask the students to write the date of their penny in an upper corner of their paper. Then—either in the classroom or out on the playground—have students hide their pennies (no burying, please). Then have them create a map that reveals their penny's location. No words are allowed on the map—only pictures or symbols. When all of the maps are completed, have the students swap maps and see if they can locate the hidden treasures.

Students can also use the penny treasure hunt as a way of practicing the composition of clues. Challenge them to write clues utilizing a variety of methods:

➤ body orientation—using left and right;
➤ cardinal direction—using the points on a compass;
➤ pacing—counting the number of steps; and
➤ observational clues—relying on features in the built or natural environment.

WORKING WITH A GROUP IN THE FIELD

Before we go out in the field to work on a Quest with a large group, we try to break the group or class into smaller working groups or "teams." There are two basic approaches to this division: by *task* or by *locale*. When dividing by task, each team can be assigned a different job: creating the map, drafting the clues, putting together the Quest box elements, and so on. When dividing by locale, each group composes map elements and clues for a different subsection of your Quest.

The size of the teams will vary, depending on several key factors. First and foremost, what are the characteristics of your chosen Quest site? Is your Quest in town or in the woods? Is it in a contained area—a park or cemetery—or does it meander along a trail or river? The more contained the area is, the smaller your groups can be. Another factor is the overall size of your group or class: are you working with a class of eighteen or a scout group of eight? What about your group's maturity level? Can your teams function without adult supervision? What kind of support do they need? Finally, if you are working with smaller children, you will need to consider the number of adults that are available to help as chaperones and group leaders.

Establishing Teams

If you are working in a contained Quest site—a small cemetery, for example— you might have many teams made up of pairs or trios working without direct supervision and only one adult for every four to six pairs or teams. If, however, your Quest winds through the woods, stretching the group out across a distance of a mile or more, you'll want to work with larger teams. The number of teams you end up with will be determined by the number of chaperones you have: if there are two chaperones, you'll have two teams, and so on.

Team Sizes

The ideal size for a Quest-making team is two to four people. With a group this size, everyone can be fully engaged in the task at hand. This group is also small enough for you to maintain their focus without having to continually struggle with side conversations and other distractions. A group of this size can be self-monitoring and include a wide range of skills and abilities. Having "mixed groups" composed of individuals with varying abilities and skills allows each team to successfully address the challenges that arise when working out in the field.

Interestingly, with school groups, some children who perform poorly in the classroom setting often excel in fieldwork. Again and again they prove the most observant, the most sensitive, and the most comfortable outdoors—budding foresters, historians, or artists. When working with students, it is also good to break up the "best friends," whose habits of camaraderie can get in the way of bringing their best work to the project. Teams can certainly be larger than four; however, beyond five or six it becomes appropriate to choose a team leader or include a chaperone whose main role is to help the group stay focused.

Chaperones

The more chaperones, the better, both for effective teaching and for promoting community involvement. In the best of all possible worlds, your chaperones might include:

➤ school staff, like the art teacher (for field drawing), gym coach (for orienteering), or technology coordinator (to teach digital photography);

6.4 Cornish Flat cemetery team. Photograph by Steve Glazer. During the Quest-making process, student groups will bond with their sites. Take pictures to record and strengthen these feelings.

➢ parents who can offer enthusiasm, assistance, and promotion to the project;
➢ a person who knows your *site* well—a volunteer from the historical society, the property owner, a bird watcher; and
➢ a community member who is knowledgeable about your Quest *topic* or theme.

Team Names

Once you have your teams set, you might simply refer to them as teams 1, 2, and 3 or A, B, and C. But better still, as you learn more about your site, your teams can adopt or be assigned names that link to particular site characteristics. As we worked on the Porter Cemetery Quest in Lyme, New Hampshire, student groups adopted the names of the families they were studying: one group was the Porter family, another the Perkins family, and a third the Turner family. This added richness to the Quest and helped the students to further bond with the site and become invested in their research. For example, the Perkins family was no longer simply an object, concept, or subject that the students were studying, it was an identity they had adopted.

6.5 Abraham Perkins stone. Photograph by Simon Brooks. Select names for your groups that reference site attributes, such as the "Perkins Family."

In creating other Quests, groups have adopted buildings (the school, a church, the library, a mill), trees (oak, maple, hemlock, birch), Quest elements (the overlook, the beaver lodge, the dam), and so forth. If you are assigning different tasks to each team, you might want to consider asking each team to write a "mission statement." Writing this statement will help them clarify the task at hand and the boundaries between tasks; it will give each team an understanding and respect for the work of the other teams; and, most importantly it will provide the "big picture" of how the whole thing fits together in the Quest. An example from Wendy Siden's fifth grade class: "Our responsibility is writing clues in the form of couplets that lead up to the treasure box, making sure that they are legible, make sense, and are specific so people who are doing the Quest will know where they are going."

Outdoor Classroom Ethics

As we head out into the field with students, we like to tell them that they will be visiting a larger classroom today, one that is much, much bigger than their regular classroom. "You won't believe how large this classroom is," we tell them. "You will be amazed. The walls . . . they are so far away you can't even see them. And there are wild animals hiding in this classroom—birds, squirrels, raccoons, spiders. This is a classroom that can compete with Max's room in *Where the Wild Things Are*."

That said, however, as large and as wild as this place is, it is still a classroom. And the rules that apply in your classroom should apply out in the field as well. It is best to let your class help to determine its own rules of conduct, perhaps using the Quest Site Ground Rules in chapter 7 as a starting point. Then let students remind each other of their rules in the classroom before heading out into the field, or while being driven to the site. Review the rules each time to make things perfectly clear—and so that appropriate behavior becomes a good habit.

To keep younger students' energy focused, we often develop activity worksheets: data collection forms, critter collection sheets, building inventory sheets, and the like. These forms should include student name, date, and site location. Giving the students boxes to fill in makes the task clear and helps them be clear about the work.

Name of the Day

With younger students, a nice closing ritual to each day of field study is a short session of field journaling at student's special places, followed by a reconvening at the meeting spot to select a "name of the day." Elicit from the students three candidates, then let them choose the one that best suits the day they've had, for example, "the day we found Asubah."

Back in the Classroom

As much as possible, incorporate your fieldwork into your lessons. On the Porter Cemetery Quest, fourth grade students collected information from gravestones in a small cemetery. Back in the classroom, each student was responsible for entering that data into a spreadsheet. The students then learned how to sort that data and build bar and pie charts. Those same numbers—the dates—can also be used for skill-building math problems. For example:

1791	The year of Abraham Perkins's death
−32	His age at the time of his death
?	The year he was born

Stone#	Last Name	First Name	Middle Name /Initial	Month of Birth	Day of Birth	Year of Birth	Month of Death	Day of Death	Year of Death	Age at Death	Male or Female	Husband/Wife of	Son/Daughter of
51	Bell	James		8		178?	1	7	1787	<7	Male		John Bell
7	Bell	John									Male		
57	Breck	Nathan					12	22	1854	63	Male	Hannah Chaptin	
57	Breck	Nathan		10	16	1781	12	22	1854	64	Male	Hanah Capin	
58	Buller	John					5	21	1882		Male		
70	Carpenter	Baxter				1831			1832	1	Male		Anna and Asa Carpenter
75	Carpenter	Jesse					12	20	1820	1.8	Male		Asa and Anna Carpenter
76	Carpenter	Anna					2	24	1813	4	Female		Asa and Anna carpinder
66	Carpenter	Anna					5	25	1843	56	Female	Asa Carpenter	
	Carpenter	Asa					9	10	1862	78	Male		
29		Selly				1845			1845	0.34	Female		William L. and Sally M.
9	Heaton	Asubah									Female		
22	Hebard	Mary					2	6	1876	88	Female	James Hebard	
	Henry	Gorge					1	1	1869	1.7	Male		Hunaball and Marinetas
56	Hewes	Moody					2	8	1841				
19	Hibbard	Lydia				1750	7	10	1840	90	Female	Wife of	
23	Hubard	Mary	Ann			1811	2	21	1846	35	Female		
26	Michell						9	20	1820		Male		J.L and Betsy Michell
60		Sophronia					11	3	1870	66	Female	Asa Carpenter	
59	Packard	Nancy					4	18	1797	2	Female		Jairus Packard
73	Packard	Lucy				1788	12	1	1811	33	Female	Jurius Packard	
34	Paine	Hannah					10	8	1828	39	Female	William Paine	
28	Perkins	Abraham		6	14	1791			1827	36	Male	Fear Perkins	
24	Perkins	Fear					11		1796	39	Female	Abraham Perkins	
5	Porter	Deacon	Cavin				8	14	1831	81	Male		
1	Porter	Thomas							1808		Male		
30	Smith	Theodotia				1808	1878			70	Female	David Tallman	
31	Tallman	Thomas					1	24	1849	80	Male	Hannah Tallman	
32	Tallman	Hannah					3	7	1891	92.1	Female		
30	Tallman	David				1822	4	20	1880	68	Male		
18	Turner	Rhoda					4	15	1850		Female		
14	Turner	Jacob					1	22	1804	54	Male		
10	Turner	David							1826		Male		
19	Turner	twins									F&M		
38	Wadleigh	Harriet	N			1801	8	5	1814	13	Female		
38	Wadleigh	Nancy					5	3	1841	67	Female		
	Wadleigh	Benjamin									Male		
36	Wadley	Henry	D				6	19	1851		Male		
25				8	27	1802	8	29	1802	0.01	Male		
34		Hannah					10	8	1828	39	Female	William Paine	

Bell — James — John   Heaton — asubah   Henrey — George   Michell — ???

Hebard — Mary   Hews   Packard

Breck — Nathan   Carpenter — Baxter — Jesse — anna — anna — asa   moody   Hilard — Lydia — Halried — mary

6.6 Cemetery records created by fourth graders at the Lyme School. Questing students often collect data on the site. Back in the classroom, data can be entered into a spreadsheet. During the Porter Cemetery Quest, students collected names, spouse names, and dates of birth and death from the headstones of a small cemetery.

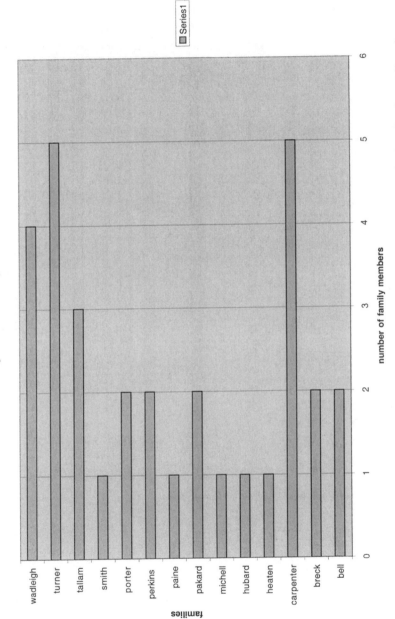

6.7 Cemetery graphs created by fourth graders at the Lyme School. Older students can sort the data they have collected, create charts and graphs, and develop hypotheses that interpret the data.

THE SEASONS OF A QUEST

After two visits, your students will already know a lot about the Quest site and be ready to make use of primary and secondary sources. After three visits, they will have enough curiosity and knowledge to benefit from a community elder or resouce person. By this point, you've hit the "peak" in your Questmaking, and it is time to "bring it all home." We usually focus on establishing the overall route first, then the clues, the map, and the treasure box components. When you are finished, remember to celebrate your success and publish and distribute the Quest so that others can learn from all of your hard work!

7) Out in the Field

BY NOW YOU HAVE PICKED your own Quest site. Perhaps it is the wood lot behind the school, the wetland at the end of a cul-de-sac, or a row of technicolor Victorian houses downtown. What comes next? Seeing what lies out there, waiting for you and your group, in the field. Walt Whitman, in *Leaves of Grass*, wrote:

> Stop this day and night with me and you shall possess the origin of all
>   poems,
> You shall possess the good of the earth and sun, (there are millions of
>   suns left,)
> You shall no longer take things at second or third hand, nor look
>   through the eyes of the dead, nor feed on the spectres in books,
> You shall not look through my eyes either, nor take things from me,
> You shall listen to all sides and filter them from your self.[1]

We, too, can possess "the origin of all poems" by becoming completely immersed in where we are and in touch with the situation unfolding around us at this very moment. This is the work of being out in the field: not taking things at second or third hand, not looking through the eyes of others, but discovering what is there through the filter of ourselves.

WORKING WITH YOUR GROUP ON LOCATION

In the field, a sense of possibility pervades: perhaps it is a feeling of richness, a sense that our communities are so much more than we've imagined. Perhaps we have the feeling of history coming alive, or a feeling of magic, or an experience of beauty. Whatever the feelings, they can be nurtured by approaching fieldwork through a deliberate, creative process.

The first step in this process is establishing a "gate," a transitional passage point where you move out of your ordinary world and perception and into the Quest.

Establishing the Gate

Most likely, each time you visit your chosen Quest site, you will park and begin your work in the same location. Whether working alone or with a group, it is helpful to establish, for the duration of your project, a spot where you can gather. You can use this place to go over instructions or activities, answer questions, share readings, or meet special guests who may be joining you for the day. A good Quest might include many visitors over the duration of the project. Your gate might be under a shady oak, along a gently sloping hillside, or on the steps to a public building. In Cornish, New Hampshire, we would gather on a large boulder. In Fairlee, Vermont, we met in the boarded-up doorway of the old depot.

As you arrive at your Quest site for the first time, look around with your group and try to find an easily accessible spot, no more than a minute or two away, that can both accommodate and contain your group for a few minutes. The quality to look for is that of an amphitheater—an arrangement that is protected on at least one side to help create a central focus. This meeting site will help you to establish a routine and create a jumping-off point for what comes next.

Quest Site Ground Rules

These simple ground rules will allow all of your other activities to proceed smoothly.

- Be sure to allow yourself (or your group) plenty of time.
- Be prepared for weather.
- Keep to established paths.
- Avoid damaging fences and stone walls.
- Do not disturb or remove cultural or historical artifacts.
- Observe wildlife from a distance.
- Do not feed wild animals.
- Don't litter—pack it in and pack it out!
- Carry sufficient drinking water.
- Respect wildlife, plants, insects, and trees.
- Respect the property of others.
- Be careful in wet areas, on steep slopes, and in sensitive habitats.
- Always have a good map and compass whenever you go into an unfamiliar area—as well as a field guide and first aid kit.

➤ Be respectful of other beings and the land.

➤ Have a wonderful time while out Questing!

Inspired Readings

Share inspired readings each time you gather at the gate. Choose readings that relate to the particulars and specifics of your Quest site. While working on the Lyme Sheep Quest, for example, we would read poems like Robert Frost's "Mending Walls" or "Bending Birches." Another poem we have used successfully is "How to See Deer" by Philip Booth:

> Forget roadside crossing.
> Go nowhere with guns.
> Go elsewhere your own way,
>
> lonely and wanting. Or
> stay and be early:
> next to deep woods
>
> inhabit old orchards.
> All clearings promise.
> Sunrise is good,
>
> and fog before sun.
> Expect nothing always;
> find your luck slowly.
>
> Wait out the windfall.
> Take your good time
> to learn to read ferns;
>
> make like a turtle:
> downhill toward slow water.
> Instructed by heron,
>
> drink the pure silence.
> Be compassed by wind.
> If you quiver like aspen
>
> trust your quick nature:
> let your ear teach you
> which way to listen.

You've come to assume
protective color; now
colors reform to

new shapes in your eye.
You've learned by now
to wait without waiting;

as if it were dusk
look into light falling:
in deep relief

things even out. Be
careless of nothing. See
what you see.[2]

From a Questing point of view, this poem is wonderful. It makes perfectly clear a number of key points:

Go . . . lonely and wanting. Or stay and be early. Every place is different at different times of day, or with the changing seasons. It pays off to visit your site early in the red sky of the new day—or to linger in the falling dusk.

Expect nothing always; find your luck slowly. If we are too full of expectation, our focus can become so narrow that we lose our openness to the many things that are actually happening right now, around us. Seeing, like learning, takes time. In observation and experience, slower is better.

Learn to read ferns. Moving quickly through the world, everything bleeds into one: a tree is a tree, a house is a house, and a car is a car. The world seems generic. But this, of course, is not true. The world is filled with precise details. Looking closely, even ferns are filled with a depth of variety, mathematical pattern, artful symmetry, structural complexity, and tangible meaning that boggles the mind. The presence of a specific fern might betray a soil type, a level of moisture, the relative amount of light or shade, or the type of forest community.

Drink the pure silence. Silence is the space that allows the perception of sound and the witnessing of movement and stillness.

Be compassed by the wind, and *Let your ear teach you which way to listen.* Trust your senses. A dazzling array of information is approaching you and passing through you at this very moment by way of your senses of sight, touch, hearing, and smell. Open up to this wisdom.

Be careless of nothing. See what you see. Don't see what we see. What do you see?

This poem sets the tone and can help your group "listen to all sides and filter them from yourself" or disappear like Taoists into the mists of a Chinese painting, becoming part of the landscape.

Special Spots

After opening our first meeting, we ask participants to wander silently and aimlessly for a few minutes so that they may discover a "special spot." Emphasize the *aimless* quality. Aimlessness allows what at first appears to be background to emerge into focus as foreground. Given a little time and space— being aimless and goalless—the surroundings beckon us, each in our own way, to a place of comfort and belonging: this small flowering tree because it is precise; that building because it has a heavy, noble presence. Once we find them, these special spots are places we can return to on subsequent visits for five or ten minutes of silent observation, sketching, and journaling. Time spent at special spots serves a number of important purposes:

> it provides an opportunity for group members to bring down their energy and become more focused;
> it allows us to become more aware of our surroundings; and
> it offers us the opportunity to hone our observation, writing, drawing, and record-keeping skills.

Three ground rules for special spots: this is silent time; everyone must be close enough that they can maintain either eye or sound contact with the group leader (no straying too far); and individuals should keep at least ten steps apart, so that each can focus on a place of their own.

During each site visit we hope for at least one page of journal entry. We also ask that each journal entry be noted with today's date/site location/time/ weather.

Places change over time, certain events may be site specific or time specific, and the weather affects both the things we see and how we feel. It is important for these elements to be a part of the record-keeping process. Joseph Cornell's book *Sharing Nature with Children* offers more ideas about how to use special spots.

7.1 Journal writing at a special spot. Photograph by James E. Sheridan. When working with a group, we recommend that every individual choose a special spot that they can return to on subsequent visits. Special spots can support a wide array of reading, writing, artistic, and observational activities.

Special Spot Journal Prompts

Some groups do better with free writing, while other groups require structured prompts. There are many fine books filled with journaling prompts: Natalie Goldberg's *Writing Down the Bones; A Crow Doesn't Need a Shadow* by Lorraine Ferra; *A Life in Hand* by Hannah Hinchman; and Susan G. Wooldridge's *Poemcrazy*, among others. Here are a few simple, favorite journaling prompts:

Notice What You Notice. Make a list of everything that you sense. For example:

> White chop of waves on the harbor.
> The words "Thank you" on the trash receptacle to my right.
> Peeling brown paint on the seat of my bench.
> The hum of the ferry engine in the background.
> The scratching of my pencil in my notebook.

Notice how this particular list moves from visual to aural, and from further away to closer in? This exercise was taught by poet Allen Ginsberg during his summer writing programs at the Naropa Institute in Boulder, Colorado.

Sound Map. Find a place. Sit quietly for a few moments. Then:

1. Draw a circle—that is the horizon.
2. Draw a dot in the center—that is you in the middle.
3. At the top of the circle—where twelve o'clock would be on a watch face—draw a very rough sketch of whatever is right in front of you.
4. Then sit in silence.
5. As you hear sounds, try to mark them on your sound map, using symbols. Sounds behind you and to the left might go at seven or eight o'clock. Sounds in front and to the right might be at one or two o'clock. (Sound maps move us away from thinking into listening.)

Name Poem. Compose a poem on your site and about your site using the letters of your name, as we have done here.

> Stretching high above me
> Trees
> Evergreens
> Very green white pines
> Easing back and forth in the
> Noon breeze.

S O U N D   M A P

Instructions:

1. Sit in silence.
2. Use symbols   sketches   or words   to describe what you hear.
3. Mark on this map the sounds that you hear where you hear them:

  - the sounds <u>in front of you</u> go above the circled you
  - the sounds <u>behind you</u> go below the circled you
  - the sounds <u>to the left</u> go to the left
  - the sounds <u>to the right</u> go to the right

7.2 Use a sound map as a way of recording the sounds of your site. Sound maps work equally well in urban and natural settings.

> *D*aybreak,
> *E*arly morning sun
> *L*ast year's crumbling leaves
> *I*n the fragrant spring
> *A*pple blossoms.

The lines do not have to be equal in length. This allows more emphasis to be on the observation as opposed to the structure of the poem.

Drawing in the Details. Pick up a small object, a pine cone, for example.

1. Draw it.
2. Make a list of words to describe it.
3. Ask questions about it.
4. Invent hypotheses about it.

Your group is now at the site in a quieter and more observational and sympathetic mode, ready to move into site exploration.

SITE EXPLORATION

Every place is unique, and each place is filled with elements that can be put together to tell a story. But to find that story, you need to pay attention. First, close your eyes. Be silent. Breathe. And listen. Start with absence and darkness, and then open your eyes. Notice: is anything moving? Is anything making noise here? What do you smell? Each sense serves as a doorway.

Sound: Coming from behind a wonderful old tree is a humming noise. What can it be? "If there is a buzzing noise, there must be something making a buzzing noise," said Pooh Bear. Is it the power lines? The hum of a transformer? No, it is the sound of the highway—Interstate 91—marching north and south, two lanes in each direction, through the Connecticut River Valley. We can *hear* the highway but can't quite *see* it. Are there other sounds? The sound of the wind passing through a majestic tree. And what kind of tree is it?

Sight: Let's look at the overall structure of the tree, then underneath the tree, and then at its leaves. The tree has wide, stretching limbs. The breeze has dropped round, lobed leaves and acorns to the ground. Acorns are a clue that this is an oak—but what kind? A field guide to trees lets us know that this is a white oak. The other oaks in this part of the country have sharp, jagged lobes.

Now let's try smell. We breathe deep. What do we smell? The smell of freshly mowed grass? The smell of manure laid down upon a field? The smell of the highway? Here, the smell was of diesel exhaust.

Gather your group, backs facing the center of a circle, like a barrel. Close your eyes—all of you—and then open them anew, in silence. What do you see, hear, smell? Ask participants to call out, one at a time, the name of things that interest them.

The first entry point into your site is through direct perception. Once you have identified a few interesting features, you can split the group into teams and let them explore these new microhabitats, making general notes and drawings, and recording questions that arise in their field journals. Regroup, move to the next vantage point, and repeat the process.

At the close of your first session, brainstorm the elements of the site that people found most captivating. Even after a single visit, eyes will already have opened to the richness that lies around them.

7.3 White oak. Photograph by James E. Sheridan. What do each of your senses tell you about this tree?

Possible Storylines

In one short site visit to the Ely Cemetery, a number of possible Quest storylines arose:

➤ A story about the cemetery in general (which was cut off from the village by the building of the interstate).
➤ A story about Ely Village—which includes the cemetery but also fans out to include things like the village store, post office, train depot, and so forth.
➤ A story about one (or more) families interred in the cemetery.
➤ A story about the beautiful white oak tree, starting or ending beneath its broad canopy.
➤ A story focusing on trees that includes the white oak as a stop.

REVISITING YOUR SITE

As you revisit the site, even the second time—and certainly by the third visit— its key places and how they fit together into a narrative are beginning to take shape. That narrative will begin to dictate the field activities. Do you need to measure trees? Draw portraits of buildings? Collect the numbers of street addresses? Collect data on and photographs of tombstones? Inventory the residents of your chosen stream or pond? Copy the exact wording from signs to use for "letter clues"? Compose directions that help someone wander through town, following precisely in your footsteps? Draw and describe features and milestones along a path?

It just so happens, as you move down the road from the Ely Cemetery, you pass a nice variety of northern forest trees, including a few remarkable specimens: a large white ash; some large hemlocks; an ancient, tattered yellow birch. Because of these trees, this Quest became the Miraculous Tree Quest.

We structured the Quest so that it would move from tree to tree. Our goals with the finished Quest were to teach the basics of tree identification and to focus on the historic relationship among people and trees: how we have used the forest to meet our needs and how our actions have modified the forest, for better or worse. The Miraculous Tree Quest, taxonomically, falls into category A-4 as a "number hunt" (see chapter 5). The numbers are earned by measuring the diameter at breast height of the different trees you meet on the Quest. This Quest requires a tape measure.

The Miraculous Tree Quest

I stand in back with wide, stretching arms,
Protecting all here from harm.
My family is oak, my color white—
A compatriot stands next to me on the right.
Rounded, lobed leaves can be found
Either upon me—or upon the ground.
Please measure my waist, up four feet,
In order to calculate the DBH—hey, that's neat!*

Record the white oak's DBH here: _____

Leaving the highway noise behind,
Seven feet up on Allbee a finger pointing up find.
Then a baseball diamond sitting on home
Slink under the line, and left up the road roam!
We're now in a mast-producing zone—
A fancy way of saying nuts call this place home.

*Foresters use a standard measurement for the girth of trees, known as the "diameter at breast height" or DBH. On this Quest we will collect the DBHs of a number of trees. To find the DBH, measure the tree's diameter approximately four feet above the ground.

Curving to the right,
Look through beech on your left.
A road meanders lower, tracing the ravine
Where during the Pleistocene[†]
The Ely River could be seen.[‡]

The road curves right past a pole
And the tall, twin hemlocks.
Hemlocks are one of the four
Dominant northern forest trees
Growing up here among the rocks.

Measure the first hemlock's DBH: _____

Onward! Sugar maple, beech,
And yellow birch are the other three—
Oh say how many can you see?
Curve left. Thread bare from the years,
An old yellow birch stands right near 675
Looking ragged—yet it is still alive.

Calculate its DBH: _____

Our fork holds two poles in its crest—
Veering left I think is the best.
The next big tree on the right is a white ash.
How do we know?
Like a muskmelon its bark does grow!
What do folks do with a tree like that?
Make hockey sticks or a baseball bat.

The ash's DBH: _____

You'll pass a sugar shack on the left,
And a frontyard spruce on the right.
But stay straight ahead
With your goal still out of sight.

[†]Better known as the "Ice Age."
[‡]The Ompompanoosuc is a descendent of the Ely River, which used to drain out through this valley, approximately tracing the path of Route 244.

An apple tree looms left flanking the road.
In fall it will bear a tasty load.
If soon you pass the numbers 369
I can tell that you are doing fine.

Then, uphill on the right
Three white birch sit tight.
Just past them is a driveway
Which we will turn up today.

Beyond the birch is a majestic tree.
Oh tell me what you think it might be?
Despite a very terrible blight,
This century old _____
Still stands living in our sight!

A miraculous tree!
A rare sight to see!
Record the miraculous tree's DBH, please: _____

Now examine the nuts, the bark, buds, and leaves—
For then you will see how to see this rare tree!

To find out the name of the tree you have found,
Add up the DBHs and then measure the ground.
Taking that distance, look 'round for a hole
Where you'll find your answer—but don't tell a soul![3]

QUEST OVERVIEW

Along the route of this Quest, you visit trees, and each stop offers the opportunity to enter into intimacy with a particular species of tree by looking closely and measuring. Differences are revealed through comparison. Learning comes through looking, touching, and a little math. Collecting those measurements, and adding them together, creates a sum. This sum is the final clue—a distance revealing the location of the Quest treasure box.

To determine the identity of the "miraculous tree," first, we examine its most distinctive features. Then, our final clue—our sum of six trees' DBHs—leads to a hollow and the treasure box. The box includes a custom-designed

7.4 So what is the "miraculous tree"? Photograph by Simon Brooks.

stamp, a stamp pad, a sign-in book, and an essay that shares the miraculous tree's story, which is reprinted here. The story is excerpted from the chapter "In the Shadow of Giants" in Ted Levin's *Blood Brook:*

At the beginning of the twentieth century, every fourth tree in the hardwood forests of the central Appalachian Mountains was a chestnut. No other nut-producing tree, not even all the species of oak or hickory collectively, had inspired as spirited a following as the chestnut which brought the flocks home to its shade. The nut crop, annual, dependable, and huge, fed wild turkeys, black bears, white-tailed deer, foxes, gray squirrels, opossums, and raccoons. People ate chestnuts, ate the animals that ate the chestnuts, and employed chestnut wood for almost everything from cradles to coffins, as the wood split with straight grain, worked easily, and was resistant to rot.

The chestnut blight arrived in the port of New York with nursery stocks of Asian Chestnut trees around the turn of the century and was first detected in 1904 on the American Chestnuts that lined the walkways of the Bronx Zoo. The blight did to the American Chestnut what smallpox did to the American Indian. Swift and lethal, it felled whole forests, nations of trees. Over the next fifty years, fungal spores were spread by wind, birds, insects, and every native chestnut stand in eastern North America had bowed under the blight as if it were an ill-wind from the Old Testament.

The origin of the Ely tree is a mystery. A chestnut pathologist from Connecticut believes that east-central Vermont is not part of the native range of the American chestnut, and therefore the tree had to have been planted sometime in the last century, a once common practice that extended the chestnut's distribution west to Wisconsin and even Oregon. The tree's current owners claim the Ely chestnut seeded in during the tenure of the previous owner, a Vermont hill farmer who told them that the tree had grown there as long as he could remember. One thing I'm sure of, this is the largest chestnut I have ever seen."[4]

This Quest leads to a mature American Chestnut tree. For many visitors, this will possibly be the only mature American Chestnut tree they ever see in their lifetimes. The Quester leaves the scene with a stamp impression, but more than that, an experience: a memory and much more knowledge about the plight of the chestnut.

Out in the field, open, looking . . . this is precisely where stories unfold and the spark of love for place can emerge. Be prepared for unplanned, teachable moments—things, as is their nature, will arise in and of themselves, in their own time. Be wide open. Witness them. Catch them. As Philip Booth reminds us, "Be careless of nothing. See what you see."

8) Researching Your Place

TO MAKE A REALLY GREAT QUEST, you need to do several different things: discover a story that is unfolding in your chosen place; break that story into a progression of smaller themes and scenes; then stitch them together again in a playful, educational way with your movement and teaching clues. How can you learn more about the story of your place? Through making observations and conducting research. Research takes many forms. There will be observations recorded first hand out in the field, visits with individuals and/or groups, and research done in the library, in community archives, in attics, or online.

Let's pretend, as an example, that you are one of the Kearsarge, New Hampshire, region homeschoolers, and you have chosen the Esther Currier Wildlife Management Area as the place to develop your Quest. You visit this new environment for the very first time, and while out in the field make some initial observations:

OBSERVATION: The site is named after Esther Currier. Who is Esther Currier? Why is this site named after her?

OBSERVATION: There are at least three beaver lodges. Two are sprouting trees, while the third one still seems occupied. How can you tell? You have seen freshly cut saplings, with a dusting of wood chips and shavings around their bases. You've seen tracks, and ruts leading down the bank and into the pond. All of these things seem to be the result of beavers gathering and hauling supplies.

OBSERVATION: You see a beaver swimming out on the pond and some turtles sunning on downed trees.

OBSERVATION: You have found (and walked across) the beaver dam that appears to be holding back the pond. (The intact dam can be interpreted as further proof that the beavers are still here and the dam is being maintained.)

OBSERVATION: You have discovered a small grove containing stumps of trees that have been gnawed down, apparently by the beaver(s). Was this to maintain the dam? For browse? How can you determine this?

OBSERVATION: You saw a large snapping turtle crossing the trail in a sunny, sandy place. Why was this awesome creature so far away from the pond?

OBSERVATION: You have discovered a soggy place that must be crossed by walking on wooden boards. What's going on here?

OBSERVATION: You have found two small huts. What are they?

You've made notes and a number of sketches in your field journals and taken a few photographs. Both the sketches and the photographs reveal that the still, reflecting water is pierced by many standing dead trees, or snags.

So these are your observations. This is a great place for a thematic Quest about how beavers modify a habitat. But how can we learn more about this place? About any place?

WHAT'S IN A NAME?

Always start with what you know. We do know the name of this place: the Esther Currier Wildlife Management area. Back at the turnoff into the parking lot, a wooden sign said as much. Closer observation of the sign, however, indicates that this property is maintained by a number of local organizations, including the Ausbon Sargent Land Preservation Trust. A call to the land trust's director, Debra Stanley, reveals that Esther Currier was a member of New London's conservation commission for twenty-two years. It was she who first identified this site's potential as a nature preserve.

Here our search for "what's in a name?" was short: the observation of a name on a sign led to a phone call and then the desired information. But the key point is to start with what you know and then try to circle outward and find others that may be able to help you answer your question. If there were no sign here, we could have tried other courses of action. We might have taken a trip to the town office and looked at a tax map. Or, after composing a precise description of the property (e.g., southeast of Route 11 in Elkins, New Hampshire, at the dirt road across from the cemetery), we could call the planning or zoning office. Other strategies might have included approaching neighbors, querying parents, or approaching longstanding community members—in this case, loggers, naturalists, hunters, or conservation commission members (the people in the community who have spent a lot of time wandering around outside).

Our call to Ms. Stanley and the Ausbon Sargent Land Trust produced more than an answer to our question, however. We were also given a wonderful site map that had already been developed. That meant we didn't have to create a Quest map from scratch. We lucked into one of the many resources that are usually out there, just waiting to be found.

USING MAPS

Maps are great resources to consult when developing a Quest. Usually you can find more than one map that includes information relevant to your Quest site. In the course of your research, it is likely that you will encounter a variety of maps. Here are a few examples to help you get a feel for what you are looking for.

Road Maps. Designed for automotive travel, these can be helpful as you create the directions to your Quest site. They generally orient your site within the context of the cardinal directions and help you understand your site in relation to the forces at play within the larger landscape (roads, neighborhoods, open space corridors). Better road maps—the DeLorme atlas, for example—also include trails, topographical elevation lines, contour shading, and common place names. By studying these maps you can learn things like the names of the bordering municipalities, the tributaries that flow through a community, and the highest and lowest points of elevation. Road maps are a fine tool for helping you and your students or group develop basic geographical literacy.

Tax Maps. Tax maps are found in the zoning or planning office. They break down the landscape of your community into parcels, each with a discrete legal description and defined boundaries. Tax maps are useful for determining two very important things: whether your Quest site is on public or private land; and who the property's steward or owner is. This is crucial information, because this is the person, ultimately, who can give you permission to make your Quest. In the best cases, these stewards will also share with you further additional information that they have about the site. Property owners can be excellent allies: generous and full of information and personal memorabilia that can change your perspective about a place completely. Imagine that you are studying an empty field and someone hands you a photograph of a magnificent barn that once stood there. Or perhaps your Quest ends at a waterfall and someone offers you a Victorian-era photograph of visitors, in suits, with parasols and hoop skirts, posing for a formal photograph in the very same place? You may have some hesitation about contacting landowners, or asking permission, but try to overcome it. Making a Quest without permission can undermine or sour

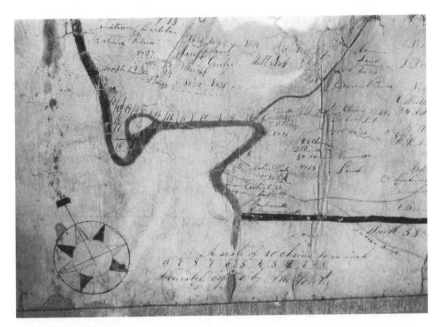

8.1 Original grant map of Piermont, New Hampshire. Photograph by Simon Brooks. Somewhere—most likely in your town, county, or state archives—you will find the first map of your place. These are fascinating documents, rich with details and stories.

a lot of good, hard work; conversely, the participation of local people in your Quest is enriching and rewarding and strengthens community.

Topographical Maps. The United States Geological Survey (USGS) publishes topographical quadrangle maps, and these maps are available for every region of the United States. They are always very helpful for understanding drainages and watersheds, and for seeing through the built environment to the underlying structure of local landforms. These maps will prove more helpful for some Quest projects than for others. USGS maps feature elevation lines, roads, watercourses, geologic features, and major infrastructure—dams, reservoirs, and the like. Using them as points of reference can help you map watersheds, student routes to and from school, and the special places in your community.

Historical Maps. An array of historical maps is also available. These can be very useful for Quests that are linked to the study of your region's history. Many New England communities still have their original hand-drawn maps, featuring the names of the initial grantees or settlers along with the name and seal of the King of England. In studying a map like this, you learn things like the

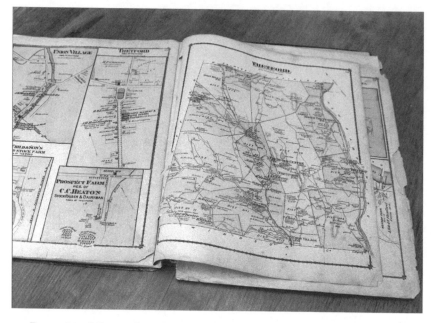

8.2 Beers atlas of Orange County, Vermont. Photograph by Simon Brooks. Village maps, common in the mid-nineteenth century, are often labeled with family and merchant names, and long-gone features from the cultural landscape: mills, blacksmith shops, wheelwrights, coopers, tanneries, or creameries.

date of your city's incorporation; who held the authority to grant title on this land; the proposed sites of the first school, town common, or meeting house; and the names of the early proprietors and settlers.

Village and County Maps. Village and county maps from the mid-nineteenth century may not always be topographically accurate; they do, however, teach us a lot about community history. Buildings figure prominently on these maps and are usually named. You will see that this was the house of so-and-so, while that factory was once a farm. Long-absent features from the landscape are prominently noted: mills, one-room schoolhouses, school districts, blacksmith shops, taverns, the local "poor farm." The larger maps often include elevation drawings of prominent structures, a business directory, and advertisements. In Vermont and New Hampshire, antique town or county maps figure prominently in Quest making. They help us see what is gone—"the ghosts"—what has stayed the same—"the venerables"—and also how much things have changed. A curricular unit focusing on the use of old maps is described in the appendix. A quick search on the internet will lead you to your state library or state historical society and a wealth of online map resources.

Aerial Photographs. While these images may not be so helpful in the development of your actual Quest, from time to time you will come across them (the internet is a good source), and they are fascinating to look at. The aerial photograph allows you to see what your chosen site looks like from an entirely different perspective. Things look different from way up there, and what seems big down here loses prominence as it fades into the surrounding landscape elements. If your site is rural, the aerial photograph is also useful in breaking down the landscape into different communities. Looking at these photographs, you can notice dramatic changes in flora where the soil is wet or dry; where the orientation and exposure is to the north or south; or where the land is steep, flat, open, or closed. If your site is urban, you can see the built landscape in relationship to the natural landscape forms. Aerial photographs can help us move beyond individual characteristics to the bigger picture or context of a place.

Geological Maps. Geological maps are available for every state and are especially helpful if the goal of your Quest is to teach about the geology of a particular site. The Surficial Geologic Map of Vermont, for example, features the last ice age's drift sheets and ice flow directions; the maximum extent of glaciation and post-glacial lakes; and the different surface elements one finds to this very day: glacial till, moraine, sand, lake-bottom sediment, and so on. Looking at this map, it becomes completely clear why the Champlain Valley is one of the best places to farm in Vermont (the soil there is made of lake-bottom sediment rather than boulders) or why Barre is famous for its quarry (there is a large bedrock exposure there). Some geological concepts, stratification, for example, are difficult to understand. However, in particular places—a canyon or cliff face—the different layers are easily seen and experienced. These particular places are the perfect places to create geological Quests.

Local Maps. Local communities, parks, trails groups, or landowner associations often create their own maps. These can be especially helpful in making Quests. They might feature local place names or commonly used trails, and they will often contain additional interpretive information that can be folded into the content (or teaching clues) of your Quest. They also work well as the underpinning of your Quest map.

NETWORKING

Let's return to the Esther Currier Quest again. Upon reviewing the site map, we discover that not only is this beaver habitat but the "soggy place with the wooden boards" is a quaking bog. We also learn that this entire site was gouged by glaciers. The glaciers have left us something called an "esker," whatever that is. Very quickly, perhaps, we have uncovered more information than we are ready to process. The story is getting complex. Help! What can we do? Who can we turn to now?

Relax. The fact is that there are many people who can help. In this case, volunteer help might come from:

➤ the Esther Currier site manager;
➤ a person who jogs here every morning;
➤ the person at the land trust who worked to conserve this property;
➤ the hunter who has been here—year in and year out—for half a century;
➤ the town forester;
➤ any one of a number of local (or regional) naturalists, photographers, bird watchers, or fisherman . . . and the list goes on.

And so it is with every Quest—and with *your* Quest. No matter the subject, you will find that there is a community of other people who *already* know or love the precise thing you are about to discover.

It is good idea to brainstorm a list of potential resources and, as the project develops, create a resource list to document your efforts for posterity.

Collecting Insight and Oral History

In the case of the Esther Currier Quest, we once again paid homage to that axiom "start with what you know" and called on Ted Levin, a neighbor and naturalist. Ted consented to make a date to walk the land with our group, and while walking out in the field, he helped us observe and learn about:

➤ the role beavers have played in American history;
➤ how beavers modify the landscape;
➤ how beaver activity affects other species—who moves in, who moves out;
➤ how glaciers modify the landscape, and what an esker is (a snaking pile of glacial till dumped by a moving glacier);
➤ why we saw the snapping turtle where we did (it was a sunny, sandy embankment (esker), the perfect place for a snapping turtle to lay its clutch

of eggs! We also learned that the eggs will develop into male or female turtles depending upon the temperature of the soil).

Reflecting back on our visit with Ms. Stanley, students also realized that human beings modify the landscape, too. This land is conserved by the land trust, which is one of the key reasons why we were there enjoying it.

As guests share with you their insights, it is important to remember to use your field journal—or all of those good words might slip away forever! A tape recorder or a video camera is a fine addition, too. If you plan on taping a guest, be sure to:

➤ ask permission to tape them;
➤ use short thirty-minute tapes;
➤ use a new tape for each interview;
➤ clearly label the tape;
➤ note at the beginning of the tape your name, the name of the person you are interviewing, the date, and the location; and
➤ present a copy of the tape as a thank-you gift to your guest.

If possible, put a copy of the tape in a public place—the nearest public library, for instance—so others may make use of this information. Remember to thank all of your visitors and assistants.[1]

The Importance of Oral History

The importance of oral history—of connecting with living members of your community—cannot be overstated. It is a key ingredient in the successful creation of a Quest. Why? First of all, because it is helpful. There are people out there who know and would love to share precisely the information you are looking for. And these generous folks will usually donate their time. But more importantly, bringing different groups of people together to do real work builds a sense of community.

Unfortunately, there are fewer and fewer things that gather the diverse age groups of our communities together. Children gather in schools; adults gather in the workplace; and elders perhaps are alone—or gathered in an assisted living facility. But often, these groups do not come into meaningful contact with each other.

The oral history aspect of the Quest provides significant opportunities for different generations in the community to come together in a new way—as allies, and as resources for each other. A student-made Quest offers the older

8.3 Oral history gathering, Beaver Meadows schoolhouse. Photograph by James E. Sheridan. Gathering a broad cross-section of the community together to collect oral history and celebrate a place will enrich your Quest and strengthen community bonds.

generation a perfect opportunity to see children as they really are, rather than as they fear they might be. Students can learn to see elders as invaluable community resources, as "keepers of knowledge."

Variations on Oral History

For each Quest, the appropriate form of oral history and community collaboration will differ. We promise, though, that the more collaboration there is, the stronger your Quest will be, the more people will use and value the Quest, and the better the community will feel at the project's close.

If your Quest is planned as a *walking tour* through a town or city, you might be best served by speaking with one or two volunteers from the historical society, the tourist information bureau, or the downtown business association. These conversations might give your group access to an interview in an important private space such as the city hall chambers or the conference room of the mayor.

A *village* Quest might be best served by an interview with a panel of community members held in one of the village's most important historic buildings: a chapel, school, or private home.

If your Quest focuses on *natural history*, the oral history interviews may take place on a college campus, in a scientist's lab, through an online chat room, or with a community forester out on your Quest site.

Regardless of the forum you choose, five key points to remember when calling on community members are:

1. *Do your homework* to make sure you invite the right people. Take into consideration that some elders may have difficulty hearing, or that others may have helpful knowledge but are not interested in interacting with younger children or students.

2. *Extend your invitation in a timely way* and make perfectly clear what it is you are asking for, that is, a fifteen-minute conversation with you, a cup of tea at your home, or a half hour with twenty students. Also be sure to specify the particular topic that you'd like your guests to speak on, that is, fifteen minutes about working in the mill and fifteen minutes of question and answer, and so on.

3. *Follow up with a reminder phone call.* Your distinguished guests may be very busy people. Be sure to remind them of this visit.

4. *Graciously host them.* Introduce them and put them at ease. If you are working with a large group, have both the group and your visitors wear name tags, so all of the participants can develop a relationship based on mutual recognition and respect.

5. *Remember to say thank you.* Nothing beats a genuine handmade thank you! Thoughtful and carefully made acknowledgements will be appreciated for years.

Remember that you don't have to do all of this by yourself. Your collection of oral history can draw on the efforts of student groups, the PTA, the library, a women's club, or a civic or fraternal organization.

SECONDARY SOURCES

A field guide to North American mammals offered insight into the beaver activity on the Esther Currier site. Another book, *Tracking as a Way of Seeing* by

Paul Rezendes, offered a wealth of information about beaver behavior, habitat, and signs. *Reading the Forested Landscape* by Tom Wessels had an entire chapter offering insights into how to "read" the presence of beavers in a landscape.

We suggest that adults and older students use these secondary sources directly. For our younger students, though, we find we can quite easily distill the information into clues and signs that even very young students can understand. For example, signs that a beaver is around include:

An intact dam
Active and abandoned lodges
Chewed-up stumps, still gushing sap, surrounded by wood chips
Active trails littered with saplings, leaves, bark, or wood chips
Tracks
A flooded landscape with standing snags

Signs that beavers have moved on include:

Broken dams
Abandoned lodges—perhaps with trees growing out of them
Older, dry, gray chewed stumps
A moist landscape pierced with standing snags
A wetland or field littered with snags . . . and the water is gone

We find that this information can be presented to younger students verbally or visually through excerpts, readings, and the creation of worksheets.

By allowing observation and experience first, and then, afterward, conducting follow-up research, students can have an "Ah Ha!" Suddenly a story that had been concealed from them has been revealed to them. Such an "Ah Ha!" might be:

➤ Oh . . . the beaver pond isn't a lake, it's a pond.
➤ It hasn't been here forever—it's new. Not man made but beaver made.
➤ The field was flooded—that's why all those trees are dead. They drowned!

These discoveries made on site become the key points of your Quest. They represent what you have learned—and what you can authentically teach to others.

Regional Literature

In addition to general field guides to plants, animals, towns, gardens, and so forth—which are extremely helpful—you will often find specialist's guides that help teach about the key thematic elements of your Quest. These specialty guides are usually published by the press of your state college or university. A few examples:

Field Guide to New England Barns and Farm Buildings, by Thomas Durant Visser. This book traces changes in barn design and construction over the course of two hundred years and teaches how to date barn eras by reading clues in materials, hardware, layout, and construction technique.

A Building History of Northern New England, by James L. Garvin. Describes the production of materials used to build northern New England structures and traces the stylistic evolution of this region's buildings from the 1700s through World War II.

New England Forests Through Time, by David R. Foster and John F. O'Keefe. The Harvard forest dioramas offer a pictorial, historical series that depicts the changes in the New England landscape over the past three hundred years by focusing on the history of one particular location.

New England Natives: A Celebration of People and Trees, by Sheila Connor. Connor shows us the life of trees within the context of the succession of human cultures—from Native Americans who crafted canoes out of white birch to colonists who built ships of oak and pine to industrialists who laid railroad tracks on the backs of chestnut timbers.

Stonewalls and Cellarholes: A Guide for Landowners on Historic Features and Landscapes in Vermont's Forests, by Rob Sanford, Don Huffer, and Nina Huffer. This gem of a book takes a short walk in words and images through the different kinds of archeological evidence you will find when heading off into the hills of Vermont (or anywhere in New England).

For more information about your particular region, call, write, or email your local state university press and ask them to send you a catalogue, or browse the "regional literature" shelves of your favorite bookstore. It is also good to look for regional literature, poetry, and small community publications. Well-written, content-appropriate literature, read onsite, is a key ingredient in the development of content-rich Quests.

PRIMARY SOURCES

Just as there are places—and people—out there waiting to teach you, there are primary sources waiting to be discovered too. Educator Monica Edinger writes in *Seeking History: Teaching with Primary Sources in Grades 4–6:*

> Primary sources are real stuff, and real stuff is powerful stuff. Anasazi cliff dwellings in Colorado. Civil War photographs. E. B. White's drafts for *Charlotte's Web.* An heirloom quilt. Birth certificates. All evoke actual past times and events. No matter how well written, no textbook can provide the same sense of being there, or realness, that primary sources provide. They help us in our struggles to make sense of the past. Studying these old things, we all act as historians attempting to unite them for ourselves into coherent and sensible stories of the past.[2]

The number of sheer facts in the world is overwhelming—yet any single one of them might prove to be the doorway that reveals to us the hidden network that is our shared coexistence, our mutual interdependence—this wonderful spinning world. Arthur Sze's poem "The Network" illustrates this brilliantly:

The Network

In 1861, George Hew sailed in a rowboat
from the Pearl River, China, across
the Pacific ocean to San Francisco.
He sailed alone. The photograph of him
in a museum disappeared. But, in the mind,
he is intense, vivid, alive. What is
this fact but another fact in a world
of facts, another truth in a vast network
of truths? It is a red maple leaf
flaming out at the end of its life,
revealing an incredibly rich and complex
network of branching veins. We live
in such a network: the world is opaque,
translucent, or, suddenly, lucid,
vibrant. The air is alive and hums
then. Speech is too slow to the mind.
And the mind's speech is so quick it breaks
the sound barrier and shatters glass.[3]

8.4 Old books. Photograph by Simon Brooks. In every community there are old books that speak volumes about your chosen Quest site.

Some of the primary sources we have discovered while out Questing in New England include old books, blotter books, town charters, ledgers, letters, maps, photographs, and signs.

Old Books. A line in a secondary source, *Patterns and Pieces,* a history of Lyme, New Hampshire, told us that in 1886 Lyme had sixty-six merino sheep dealers. But *Child's Grafton County Gazetteer* from 1886 took this information one step further, offering a listing of all sixty-six dealers by name. Why was this important? Today there are no sheep dealers in Lyme, while 120 years ago it was the biggest business in town. Times change. Lyme was once an agricultural community; today, residents commute to professional jobs in Hanover and Lebanon. Today Lyme is covered with forest—but what do you think the

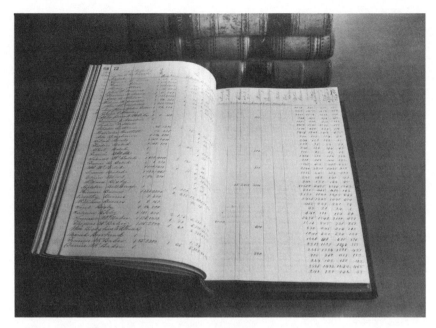

8.5 Lyme blotter book. Photograph by Simon Brooks. This book, used by Lyme tax collectors in the mid-nineteenth century, helped students to create the Lyme Sheep Quest. Counting up the entries for 1855, they came up with a total of 13,176 sheep. That's more than ten times the number of human beings living in Lyme today, and about one hundred times the current sheep tally.

town looked like in 1886? Early photographs of Lyme look like photographs of Ireland: rolling, bare hills; stone walls; huge flocks of sheep. It is wonderful to look at century-old photos where you can witness for yourself how much a landscape or community has changed.

Blotter Books. Across the street from the Lyme Elementary School is the town office. On top of one of the filing cabinets in the closet of the town clerk is a stack of oversized, leatherbound volumes. Some of these are the town's "blotter books." The town listers used to go door-to-door assessing the resident's holdings in order to determine the amount of property taxes they owed. This person had this much land, that many horses, that many cows, dogs, sheep, and so forth. With the help of simple division, you can determine the actual values of things like land and livestock. You can determine comparative values (horses vs. cows) as well. For the Lyme Sheep Quest, if you find the "blotter book" entries for 1855 and add up the total for the "sheep" column, you can determine that there were 13,176 sheep in the town of Lyme in that year.

8.6 The Risley papers. Photograph by Simon Brooks. Careful examination of old documents and letters can reveal the more elusive history: the history of ordinary people and their lives. The Risley papers collect letters, account ledgers, sketches, and epitaphs from a stonecarving family from the late eighteenth and early nineteenth centuries. *Courtesy of Dartmouth College.*

Town Charters. White pine trees were so important that they were one of the few natural elements noted by King George III in Fairlee's town charter:

> That all white and other pine trees within the said Township, fit for Masting our Royal Navy, be carefully preserved for that use, and none to be cut or felled without our special Licence for so doing first had and obtained, upon the penalty of the forfeiture of the right of such Grantee, his heirs and assigns, to us, our heirs and successors, as well as being subject to the penalty of any act or acts of parliament that now are, or hereafter shall be enacted.

Documents and Other Printed Matter. While working on the Fairlee Depot Quest, we were able to share with the students old train schedules, train tickets, a lantern, and a guided tour of the depot. The train station has not been in operation for more than twenty-five years. A cemetery Quest was enriched by the discovery of the Risley papers in Dartmouth College's special collection. These papers include drafts made for carvings and epitaphs made more than two centuries ago.

8.7 Butter box. Photograph by Simon Brooks. These Piermont butter boxes were made in a factory a block away from the school. The factory is a ghost, but a few butter boxes remain. Historical artifacts that students can actually touch deepen their appreciation for history.

Ledgers. Towns keep a ledger of accounts: what comes in (income) and what goes out (expenses). Looking through old records, you can find extraordinary details: bounties placed on predators; payments made to widows during the Civil War; the cost of everyday items, as well as the costs related to key features in the community. While making the Blacksmith Covered Bridge Quest in Cornish, New Hampshire, students were able to locate the entry for payment made to James Tasker for completing the building of this bridge back in 1881. Students also discovered that the cost of repairing the bridge a century later was roughly forty times its original cost.

Objects. An old map of Piermont, New Hampshire, showed us a rich village center where today there is an intersection. Smack dab in the middle of today's Route 10 was a butter box factory. Why butter boxes? To transport dairy products to the city. Once upon a time there was no such thing as plastic. Plastic, along with the interstate highways—both of them dependent on fossil fuels— changed the landscape of New England (and the world) forever. Only 150 yards from the Piermont Village School we found locally made, wooden Piermont butter boxes. Touching these boxes made the absence of the factory much more meaningful for the students.

Photographs. Old photographs are fascinating to children and adults alike. Every community has hundreds if not thousands of photographs that can help you discover the natural and cultural history of your community. You may find that you are granted access to originals. With today's digital technology, it is also cheap and easy for a community to safely give access to the historical record to its students and the public. All that is needed is a computer, and a scanner or a digital camera.

The Copperfield Town Quest in Vershire, Vermont, benefited from a panoramic view of a 700-foot-long coal smelter surrounded by forty or fifty buildings: houses, churches, stores, a bank, and a post office. Today not a single building remains on the site.

The Chelsea Village Quest in Chelsea, Vermont, benefited from the use of black-and-white photographs taken twenty to thirty years ago as the town prepared to submit a "Historic District" application to the Division of Historic Preservation. The photographs were used in class to help the students refine and complete drawings of their "adopted" buildings. The students also used the application's "narrative section" as they prepared biographies of the buildings.

Letters. Old letters are fascinating documents and can speak volumes about the history of your place. The George Knox Quest tells the story of one of the earliest black settlers in the community. Old documents and letters—including a letter from Joseph Reed dated March 14, 1821—were quoted verbatim in the verse of the Quest, which seeks to clarify the "legend" of George Knox:

The George Knox Quest

Community Legend has it that a black soldier
—and bodyguard of George Washington—
once lived on a hillside in the Stevens District.
Follow my clues . . . for some history and fun!

Now take this country lane back
As we travel deep into history.
To learn the story of Mr. George Knox
Pay attention to what you hear and see.

8.8 Glen Falls. Fairlee fifth and sixth graders were not the first ones to make a pilgrimage to Fairlee Glen Falls. Visitors have been coming here for more than a century. *Photograph courtesy of the Fairlee Historical Society.*

Look around.
Now once upon a time
This place was neither mowed nor paved.
And George Knox, he lived here,
With the few pennies that he had saved.

Today there are ponds, and pastures green,
But perhaps 200 years ago you'd have seen:
"1 old mare, 8 sheep, 4 lambs,
1 hog, 1 cart, 1 gig"—all of George's possessions.

A 180-year-old letter reads:
"His house is a mere hovel."
"He has lost the use of his right arm."
"He's as poor as a man can be."
"Little grows on his poor and stoney farm."

Where are these words taken from?
All these words were taken from a letter
Written and mailed to the Secretary of War
Asking that a pension be promptly sent
To help Mr. Knox, who was very old and poor.[*]

Having served his country from 1776–1777
And 1778–1781, Mr. Knox was a veteran.
As we can see in the records of his company,
Knox fought in America's revolution.[†]

Continue walking.
The driveway arcs to the right
While the stone wall stays ahead, straight.
Follow along the edge of wall and woods
To learn more of George's fate.

[*]Judge Joseph Reed wrote on March 14, 1821, to request that Knox received his veteran's benefits.
[†]Knox served in the Revolutionary War on at least three separate occasions: from December 1776 until March 1777 in the Continental Army at New York; again in 1777 in the regiment commanded by Col. Jonathan Chase of Cornish, New Hampshire; and finally in the First New Hampshire regiment from May 1778 until May 1781.

The first record of Knox in the Upper Valley
Is southeast of here, in Enfield Center.[‡]

And then later, in 1785, Knox and his wife
To James Wheelock were indentured[§]

(*Note that even today if you look at a good map*
Beyond Enfield Center you can find George Pond
And above that, a place called George Hill.
Knox Brook flows down into Lake Mascoma—
All these traces of George Knox can be seen still!*)*[‖]

Now back to indentured . . . what does that mean?
George and a wife named Peg
Were contracted by Wheelocks to serve
From 1785–1790. For each one of these five years
Twenty pounds—in land, grain, or cattle—
They would earn.[#]

Their indenture was to have ended in 1790,
But by 1788 George and new wife Jemima crossed the river.
Vermont was an independent republic from 1777–1791.
Did George cross to escape his indenture?[**]

Continue along the edge of the woods
Traversing clockwise around the small pond.
More clues regarding our friend George Knox
Can be found in stone walls you'll find just beyond.

Three stones lie still in this quiet cemetery:
One is marked "George," another "Wife Catherine,"
The third stone's words, alas, have disappeared.
What is it we can learn from this place, then?

[‡]Maurice Quinlan, p. 57.
[§]Frederick Chase, *A History of Dartmouth College and the Town of Hanover, New Hampshire.* James was the son of Eleazer Wheelock, the founder of Dartmouth College. James Wheelock and George Knox served together in 1777.
[‖]DeLorme *NH Atlas and Gazeteer*, p. 34.
[#]Charles Hughes, "Black Veteran in a White Settlement," Thetford Town Report, 1997.
[**]Town records indicate the birth of Henry Knox in 1788.

But first, let us ask and answer a few questions:
(a) Why is this family buried out here, alone?
(b) Why is a man who was so poor so well remembered?
(c) Was George Knox Washington's bodyguard?
(d) And why was it that others wrote for him his letters?

Regarding (a):
George Knox and family are buried out here, alone,
As a cemetery plot was much cheaper in the backyard, at home.

Regarding (b):
And Knox is remembered as one of the first black men
To settle here, in the Upper Valley region.

Regarding (c):
That Knox was a bodyguard we do not know . . .
But that does not prove that this wasn't so.

Regarding (d):
George Knox once used an "X" to sign his name;
Schooling, in that time, was not quite the same.

Then there is this question of—three?—wives . . .
It seems that George Knox lived multiple lives.

Looking this far back into history
A lot of what we find is a mystery:

His enlistment in 1778 says he was 32 years old.
So was George Knox born in 1746?

His pension says that he was 80 years old in 1821.
So was Knox born in 1741?

The tombstone says George died at 92 in 1825.
So was he born in 1733?

As historian Charles Hughes asked back in 1997
"What is it we can say of this black veteran?"

That:
George Knox was a man whose country he served,
And an early settler whose life should be remembered.
His life was hard: indenture, poverty, and war.
Just how much of that life could we possibly endure?

Hidden nearby, beneath some rocks
You will find this Valley Quest's hidden treasure box.

Leave us a note to tell us that you
Came to pay tribute to George Knox, too.

Last of all, look at Catherine's stone to see
That "From death's arrest no one is free."

A Quest of this depth is only possible because of the time, energy, and research of a good number of community members. However, the Quest itself
pulls research out of file drawers in Washington, D.C., and puts them into the
field for the benefit of the community and visitors.[4]

By now it should be clear: your community is rich! It is filled with resources—
places, people, expertise, secondary sources, and primary sources—that taken
together can build a strong, living curriculum that helps connect you and your
neighbors to the place you live.

9) Writing Clues

Towering high above us all
And shedding light when night time falls,
This silver rod has one green sign
Read it and you can solve our rhyme.
—The Fellsmere Ramble Quest

CLUES SERVE MULTIPLE PURPOSES

The clues your group writes for your Quest have a big job to do. In addition to telling the story of the place, they need to offer an orientation to the directions and cautions associated with this particular Quest, move Questers accurately through all of the twists and turns of the landscape, and give hints that propel visitors toward the hidden treasure box. Quite a complex set of responsibilities for a simple rhyme! Let's take a look at each of these jobs individually.

Framing the Quest

The first job the text of your Quest will have to perform is to orient people to the Quest and give them special directions, cautions, and thoughts on preparations, if any are needed. These initial clues all arrive as an overview that considers the entire journey, and with that knowledge in mind they offer strategy or advice. For example:

Welcome! This is the Trustom Pond Quest
But please remember, you're the guest.
The animals and plants who live within
Are the ones whose home you're in.
So please walk softly, don't run or shout,
But look and listen for who's about.
And please bring a pencil so you can write
The answers to the questions on the page's right.
—From the Trustom Pond NWR Otter Point Quest

Please remember to put every clue back—
Or this Quest will get thrown quite out of whack!
—From the Peabody Library Quest

Movement Clues

A Quest's clues bear the primary responsibility for moving people through town or countryside from start to finish without getting lost, and for stopping them periodically along the way. This burden can also be shifted to the map in some Quest designs. Here are a few examples of how Questmakers have used clues to lead people through the Quest:

As you stand at the boulder, "Oh say can you see?"
Head for the banner that inspired Francis Scott Key.
—From the Walk-About Quest at Frosty Drew

The ol' Vermont sugar maker says,
"Wind from the east, sap runs the least.
Wind from the west, sap runs the best."
It's a bad sap day so turn that way.
—From the Woodstock CWM Quest

By conveying basic directions in a playful and imaginative way, these clues appeal to the Quester's sense of fun and draw them into the challenge. The "ol' Vermont sugar maker" clue is strong in that it collapses into one clue both movement and teaching strategies.

Riddles and Hints for Finding the Quest Box

Another obvious responsibility of your Quest's clues is to artfully and subtly give people just enough information to find the box. The visitors may be led straight to the box, or they may be asked to collect letters, numbers, or visual clues that enable them to solve a puzzle. For example:

This small sanctuary honors a grower of plants
"And its presence is enough to make the heart dance!
A poem is tucked amidst these growing things,
One three-letter word describes a most divine being.
The third letter of this word is the one that you want
Before you head off to continue your jaunt.
—From the Lyman Point Quest

Cross over some gravel and come to a shrine
Put up for a man ahead of his time,
Who saw in these woods a great place for skiing
And hiking and learning and tree-type-I.D.-ing.
You're standing right now in a place he thought wonderful:
The site of the Dartmouth '52 Winter Carnival.
There wasn't much snow in yon Hanover that year.
College skiers came here, with their dates and their gear,
To race down this hillside where you stand today.
Moosi was among the first ski hills in the U.S. of A.
(The plaque lists two pairs of dates.
Add the last digit from each of these four dates,
multiply the total by 3, subtract 2
and write the number here: __.)
—From the Moosilauke Historical Quest

Storytelling Clues

The story clues make all the wonderful places the Quest visits come alive when the Quester's curiosity is most piqued, and they spin the Quest into a golden learning opportunity. Story clues may focus on teaching specific interrelated facts or launch into a fullblown saga. Either way, they solidify the theme and enrich the Quest experience:

I am a raptor with a snowy white head.
Banning the pesticide DDT helped me get ahead.
I build a large nest, weighing up to a ton.
Using my 2″ talons to catch fish is lots of fun.
(Look at their feet and their heads.
What 3 things make them successful hunters?)
—From the Stone Zoo Quest

The Quest is over when rock walls appear.
"Ledges Pass" they are called; the end is near,
Two gangs of workers, forty men, they say,
Took a year to carve this cut, with nary a stay.
Gas or steam engines were not yet the rule,
Hand drills and explosives their only tools.
Upon these walls are etched names and dates,
By workers whose sweat this passage did make.
—From the Enfield Rail Trail Quest

THE WRITING PROCESS

The writing of successful Quest clues involves four things:

➢ a clear sense of the purpose, theme, and story of the particular Quest;
➢ sufficient time for field observation and research;
➢ honed skills in the areas of poetry writing and riddlemaking; and
➢ a commitment to impartial Quest-testing and ruthless editing.

Let's pause to step back from the mission and content of your Quest and focus now on the writing process.

We are fully aware that a question that occurs to most people at some point, usually early on in their process of creating a Quest, is "Why the poetry?" Though one could argue that there is some doggerel, some bad poetry, maybe even some really bad poetry, within the body of existing Quests, we still maintain that it's generally worth giving it a shot. Poems are inherently so pleasurable to read, and the act of writing them brings about such focus, that we think that they should be the starting assumption in writing a Quest. We maintain that a choice to go with a prose Quest should be made consciously and for a reason, rather than out of fear, shyness, lack of inspiration, or—dare we say it?—laziness.

Poems come in many shapes and sizes, including those with strong traditional meter and rhyming patterns, those with other formal structures such as Haiku, and those in free verse, where the structure is more internal to each poem. All have in common careful attention to language—to how words sound and to what they mean. They also have in common an intention to communicate ideas crisply and evocatively—to bring the reader to a new vision or a new understanding.

Community members compose their Quest's clues in a process of exploration, digestion, expression, and then, finally, compression. Putting the clues into verse adds another layer of difficulty and complexity. However, as group members struggle with the verse, their learning about the place becomes digested, embodied, and committed to memory in a way that prose often can't seem to provide.

Our original orientation toward the use of poetry in creating Quests can be traced back to Questing's strong roots in the British tradition of Letterboxing. The British enthusiasm for metered verse seems to know no bounds, and we found it contagious. As we have moved forward in our work with schools and communities, however, we have found that our commitment to poetry with rhyme and meter has remained strong. Why? Because of its wonderfully appropriate playful qualities; because of its usefulness in school curricula—poetry and the content behind the verse is usually required by state educational standards; and because of the delightful final product, which seems to have universal appeal to potential Questers. While prose, haiku, and other forms have a secure place in Questing, we find ourselves returning often to rhythm and rhyme.

We have found that there is a tremendous range of natural poetic aptitude among prospective Quest writers. Likewise, we have found that, left unnurtured, this range in aptitude results in a wide range of quality of the final clues. Happily, some of the basic skills of rhythm and rhyme can be taught easily.

Rhythm

As any reader of poetry written by children knows, rhythm seems to be a more difficult skill to acquire than rhyme. Yet we feel that polishing an even and pleasing rhythm in your poems is worth the effort, as rhythm is a powerful and pleasurable force that can engage the readers and carry them through the poem to its conclusion.

We have found that the best way to teach poetic rhythm is to study some rhythmic poetry that works. Read aloud rhyming poetry you like and, if your inhibitions allow, go so far as to try clapping the beat to develop a kinesthetic understanding of the way the rhythm fits with the words. The poems of Shel Silverstein are wonderful for this, as are the poems of Poe, Frost, Longfellow, Kipling, and Mother Goose.

You'll find that one rhythm that is familiar to most is a series of double beats with a lighter emphasis on the first beat and a heavier emphasis on the second beat (an *iambic* meter) as illustrated by Joyce Kilmer's lines, "I think that I shall never see / A poem as lovely as a tree."

Whether you use iambic meter or another one, the idea is to find a meter that is regular and pleasing without sounding overly rigid, like a metronome. Once set, the rhythm of a poem carries the reader through the verse, but certainly, within an overall regular rhythm, variations and irregularities can give your Quest poem life.

The number of double or triple beats that you put together determines the length of your line. Pentameter (five sets of double beats) is the most common line length, since it's just about how much the average person can say in one breath. Another common line length is the tetrameter, or four sets of double beats, as in the example above. While the great masters of poetry occasionally vary line length to good effect, we suggest that, as least to begin with, you choose one line length and use it consistently throughout your Quest. Test it

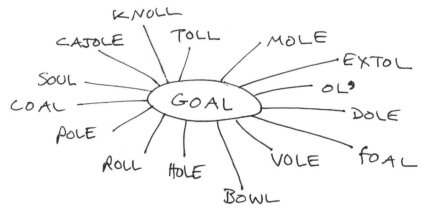

9.1 Making a word web can help you find a rhyme.

as you go along by tapping it out if necessary. We have included examples throughout this book of Quests written by amateurs, many of them children. Some demonstrate consistent rhythm and others don't. As you read through them, notice how much greater is your pleasure in those that do.

Rhyme

Most people know how to make a rhyme, but they often need some loosening up to get their rhyming juices flowing. One good warm-up exercise is to play a fast-paced round of "Hot Potato—Not Tomato." Grab a ball, a rolled sock, a rubber chicken, or something similar and assemble your group in a circle. Shout out a word as you toss your chosen object to someone across the circle. They have to quickly catch the ball and shout out a rhyming word as they toss the ball to someone else. The game continues until somebody can't think of a rhyme. They then shout out a new word and the game continues with players rhyming to that one. Time how long your group can go on one rhyme, or see how many words you can average per rhyme over a thirty-second period, and then see if you've improved after several days of Quest writing.

Another approach is to try asking group members to name something they really like—hockey, pizza, their dog, coffee. Choose one of the answers and ask the group member who suggested it to name their favorite part of that thing—the goals, pepperoni, his paws, the buzz. Choose one of these ideas and write it in the middle of a piece of paper. Then make a "word web" by writing lots of rhyming words all around the outside of the center word. We've found that there's something magical about webbing that seems to work better than listing to unlock people's rigid thinking and start the connections flowing.

It is useful to point out to aspiring clue writers that their rhymes will be most effective if they fit them into a particular rhyming scheme and then stick to it throughout their Quest. It's easiest to write the couplet, in which paired lines rhyme with each other (aa bb cc in poetic notation). A string of couplets do not need to rhyme with each other (though they should share the same rhythm), so the writer only needs to think of two rhyming words at a time. This set of couplets from the Beaver Meadow Quest is a good example:

> Turn 'round to read a big tall grave,
> For Parkhurst who was also brave.
> Proceed to the very top of the mound,
> Find the Sawyer family burying ground.
> Deborah was the first of Conant's wives,
> Find four others who lost their lives.

Assign your Questmaking collaborators to bring rhyming poems to your next gathering and have some fun studying their patterns. Others you are likely to see include:

Tercet or triplet	aaa bbb ccc ddd etc.
Quatrain	abab cdcd etc.
Limerick	aabba

It's up to you whether you make a unilateral pronouncement or a group decision about which rhyming pattern is best suited to your theme, but we strongly suggest that you decide on something and stick with it. When you cluster several lines together, it is called a stanza. The way that you cluster and break up your lines into stanzas should make sense both the rhyming pattern and the story line. Think of them as paragraphs.

Sometimes stretching for a rhyme can have a cute effect, like these two couplets from the Croydon's Past Quest:

> To the right of the dumpster you'll find Forehand Road,
> Made up mostly of dirt, and maybe a toad.

You'll hike by four turnoffs that are on your right.
Do not get too lost; there's a good chance you might.

But sometimes it's a little over the top and should probably be avoided, as in these anonymous examples:

Parallel to the stream you're walking now
Until you spot the dam and the waterfall—wow!
Enter the woods and walk upstream.
View the falls from above and below—it's a scream!

Go to the bank
and don't get spanked.

Prose, Haiku, Free Verse, and Other Forms

There are myriad other possible choices for the form of your Quest. You may have a reason for starting with a particular form—for example, you are a member of a Shakespearean study circle that has decided to create a Sonnet Quest tribute to one of your retiring members—or you might want the place and the story to dictate the form.

For a peaceful, natural setting you might want to consider the Japanese haiku, in which a mere seventeen syllables are divided into three lines of five, seven, and five syllables to convey a mood through description of a specific place and moment, usually in the natural world. Ideally, your haiku will be constructed with two parts, either one line and then two lines, or two lines and then one line, with a "hinge" between them that makes a leap, change, or "Ah Ha!" as modeled in these excerpts from the Gile Mountain Hawk and Haiku Quest:

Stop at tiny bridge:
Sensitive fern spore cases.
Each spore a new life.

Wires marching south.
Kestral perch on them waiting
For the grasshoppers.

For a contemporary story in an urban neighborhood, you might want to consider blank verse, in which there is consistent rhythm but no rhyme, or free verse, in which there is no set rhythm or rhyming pattern. In free verse, word choice and structure are integral to the message of the poem, as in this excerpt from the Trail of Silk Quest:

> Wandering in her footsteps,
> still warm?
> Find the perfect place,
> a bench,
> her bench,
> to contemplate the wharfside
> comings
> and goings.

For complex directions and/or for a content-rich story, you might choose to go with prose, as in this excerpt from the Phillips Conservation Property Eco-Quest:

> From this point you can notice more clearly the old growth forest; many large hemlocks dominate this site by blocking out sunlight with their large canopies. Look hard here and you will discover a natural gift, a young birch in the shape of a seat. The tree apparently grew upon the roots of another tree that has since decayed, leaving this interesting root growth form. Notice that the root system exists largely above ground.

Voice/Perspective

You have a choice of many different voices and perspectives when writing your Quest clues. We've found that the three voices most commonly used by young Questmakers are the "Invisible Teacher," the "Cheerleader," and the "Talking Inanimate Object," often in combination with each other. Let's look at examples of each of these. You'll easily guess which is which.

> An alder tree will be on your left as you look at the view.
> On the right are tall willow trees, of which there are two.
> Both of these tree species can be found by brooks.
> They need soil this moist to keep their good looks.
> —From the Mink Brook Quest

> Turn right down the steep, deep slope.
> Boy, oh boy! We hope you can cope!
> —From the Quechee Gorge Quest

I've got scalloped shingles on my second floor,
At School Street you'll find me at number one-four.
I'm fancied up, painted in pink and green,
Look a little bit closer—something odd can be seen!
—From the Runnemede School Quest

Experienced Quest creators have made good use of these voices as well and in addition often choose more sophisticated approaches. You might, for example, consider assuming the voice of a particular individual, such as the hero or villain featured in your Quest. Or how about speaking as an ancient saga teller or assuming the voice of a local animal? You might try taking the inanimate voice to new heights and becoming the voice of your local mountain or of the river that flows through your city or town. For inspiration, try reading through a stack of children's books and see which voices and perspectives you find most engaging. Decide upon the strategy that will best help you to achieve your desired ends, and if you are working with a group, be sure to come to agreement so that contributors share a common voice.

TRICKS OF THE TRADE

While some Quest-goers in your community will be looking for a peaceful, gently educational stroll, and others will be drawn to a physical challenge like a hike or a paddle, you will also find that there is a significant group who just love the mental workout and suspense of solving a puzzle along the way. It is to these Questers that letter and number riddles appeal.

Letter Clues

There are lots of ways that you can include letter puzzles in your Quest. Sources of letters include:

➤ Selected words on signs such as this one commemorating some "unselfish" heroes mentioned in the Adam's Trails and Treasures Quest:

> 6. With William McKinley's back to your west,
> Val's Pipe and Package is at his behest.
> A plaque tells how "Alert" men have put into motion
> Fire hoses with what kind of special devotion?
> __ __ (__) __ __ __ (__) __ __
>   6a   6b

At each stop on the Valley Quest,
You'll find a date to do the rest.
Record each date on the chart,
Add four digits-you are smart!
A little more math, you're not done,
Keep on going you'll have some fun.
Numbers to letters is the test,
To find the location of the Treasure Chest!

Across the Green, look for the sign,
Take South Street, it's not hard to find.
To the spot with a cross on the top,
The most recent date is your first stop.
On to the school, a monument you'll see
Under the eagle, dates 2 and 3 will be.
The eagle's beak on top you see,
Points to Cross St., where you should be.
The fourth date, straight ahead you'll spot
Where you may have some mail to drop.
Across the Kedron Brook you'll go
Collect date #5 as you watch it flow.
Around the corner and up the hill
To the left, don't lose your will.

An eight-sided cone is on the top
 Further along is House 21, your 6th stop.
Find 13, your NOT out of luck
Cross the footbridge, don't get stuck.
In days of old this was a mill,
Go to the left, don't be still.
To the end of the street, look up in the air,
Find the Inn's weather instruments there.
Around the corner and to the right,
The Dome of Justice is right in sight.
To the end of this street you will go,
Another right and don't be slow.
At this building turn to the right,
A big granite rock is clear in sight.
Your 7th date right here you'll find,
Could be an eight, but keep three in mind.

MDCCCLXXXIII is on the top
Around the back is where you'll stop.
The birds take shelter in the tree
In their home is where I'll be!
Pull my perch, take out the square,
Return it carefully or BEWARE!

NUMBERS TO LETTERS					
A equals 1, 26 for Z...Unscramble the letters for your destiny!!					
DATES-add across		MATH		NUMBER TO LETTER	
#1		= ☐ - 5	= ☐	= ☐	
#2		= ☐ ⨯ 14	= ☐	= ☐	
#3		= ☐ - 19	= ☐	= ☐	
#4		= ☐ - 8	= ☐	= ☐	
#5		= ☐ - 4	= ☐	= ☐	
#6		= ☐ + 6	= ☐	= ☐	
#7		= ☐ + 11	= ☐	= ☐	

9.2 Numbers can be collected and used to solve puzzles.

9.3 Searching for letters in Woodstock. Photograph by Jon Gilbert Fox. Along this Quest, visitors hunt for letters on historical markers.

➤ The names of designated objects, like swing, bench, or bridge.
➤ Guessed words, such as this one from the Hay Refuge Quest:

Off in the woods it's quiet and shady,
Benches to seat the gentleman and lady,
Gaze in the pool, then see your reflection,
Then follow the path in a westerly __ __ __ __ (__) __ (__)
__ __.

➤ Numbers can be collected and then translated into letters (1 = a, 2 = b, etc.), such as in the example from an early version of Woodstock Village Green Quest B, appearing on page 147.

> A equals 1, 26 for Z. Unscramble the letters for your destiny.

➤ Find the names of things.

> This homestead was built in 1832 in Chelsea town.
> Two beautiful front-yard elm trees were cut down.
> The place you are is called __ __ __ $\frac{}{4}$ __
> __ __ __ __ __.

➤ Single letters hidden in the landscape. These might be painted on blocks of wood that are attached to a tether and tucked out of sight, or they might be attractively printed, laminated, and displayed with permission in shop or office windows.

Once the letters have been collected, they can be assembled into words in several different ways. One method that Questmakers have used with great success is based on the popular syndicated newspaper puzzle *Jumble*, a form of anagram. This technique involves circling the letters of the clue word that will be used in combination with letters from other clues to form the new solution word. You can moderate the difficulty of solving the anagram by either:

1. Using the letters in the order in which they're collected (easiest).
2. Designating each letter with a number that specifies its position or order in the solution and intentionally collecting them in the wrong order (more difficult).
3. Collecting them in a random order and requiring the Quester to figure out how to rearrange the letters to form a new word (most difficult).

For an added twist in these jumbles, you can have Questers collect all of the letters, then subtract a few of them and just use the remaining ones. Keep your Questers guessing!

Another technique is to line the words up next to each other, crossword style, arranged so that one particular horizontal or vertical line cutting across all of them spells out a word. Check out the word puzzles in your local newspaper and you'll find lots of variations to play with.

After the letters have been assembled into a new word or phrase, this phrase can be used in endless ways to reveal the treasure location. For example:

9.4 Collecting numbers in Woodstock. Photograph by Jon Gilbert Fox. Numbers are everywhere. Collecting them on a Quest helps to keep younger Questers engaged.

- ➤ The word names the spot near the final clue where the Quest box is hidden, such as "dinosaur exhibit" or "birdhouse."
- ➤ The word names an area where the box is hidden, such as "city hall" or "duck pond" or "Miss Adam's Diner," then further clues take you to the exact spot in that area.
- ➤ The clues spell out a password. When you say this secret word to the person behind the counter at the final destination, they will turn over the box to you. This will work in any public facility, from minimarts and downtown shops to museums and town halls, that wishes to join in the game and is willing to train its staff in what to do.
- ➤ The words are collected simply for the fun and learning and to add one more texture to the Quest. You can provide a sense of accomplishment by offering confirmation of the word in the sign-in book in the treasure box.

Number Clues

Our world is full of numbers and the range of ways you can combine them into puzzles and codes is nearly infinite. You can find numbers anywhere there are signs or things to count or even to measure. A few sources of numbers include:

➤ Near roads you'll find route signs, house numbers, and numbers on telephone poles. In historic areas, calling attention to dates posted on buildings can contribute to both the puzzle and the educational story. Along interpretive trails you'll often find numbers on the signs labeling the stops.

➤ Count the numbers of red doors, windowpanes, rails in a railing, rooster weather vanes on one building, needles per bundle on a pine tree, floors in a tall building, holes in a storm drain, steps in a staircase, or even letters in a name on a sign. Here's how the Houghton Hill Quest uses a number collected from nature:

> Continue on the trail till a hemlock tree you find.
> Its needles are green, even in wintertime.
> Each needle is really a short, thin leaf, you see.
> To help the tree shed snow and not lose water is key.
> On the underside, how many white lines can you find?
> Put the number here, if you don't mind: _____

➤ Questers can also generate their own numbers by counting their steps or by measuring objects they encounter. There is obviously some room for variation and error here, though, so use these numbers for general directions and for education rather than for specific puzzle solving.

Once collected, the numbers can be used to enhance the mystery, security, or solvability of your Quest in a wide variety of ways:

➤ The numbers can be added, subtracted, multiplied, or divided together to become one number that identifies the location of the treasure on the final treasure map. This map is, of course, covered with loads of decoy numbers. You can see an example of this in the Village Green Quest E in chapter 5.

➤ The numbers turn out to be the combination on a padlock, as in the Moosilauke Historical Quest:

> When the box is in hand and your excitement is roiled,
> Drats! You are likely to find yourself foiled
> Unless with great care this Quest you've attended
> And noted above the three numbers we recommended.
> Open the box now with pleasure and glee—
> It's as easy as the numbers one, two, and three.

➤ The numbers are used as a code, with each number representing a different letter that is then used as part of a word puzzle.

➤ The clues help the Quester to identify particular objects, and they are instructed to label these objects on the map with a certain number. At the end of the Quest, these numbers on the map can be connected in order to draw out one large number overlaying the map. This number identifies the hiding place of the treasure box, as in this version of the Fellsmere Pond Historical Quest, which leads Questers to a very convincing fake rock that was built by the staff of the local zoo and contains a hidden shelf:

> Now on the map connect the numbers in order if you please
> A large number appears, which is your set of keys!
> Turn to the rocks that are quite near
> Count to the number, then stop—is it clear?
> Use your hands to feel all around.
> We guarantee that the prize you have found.

A FEW TIPS FOR SUCCESS IN THE WRITING OF CLUES

Appeal to the Senses

There are myriad sounds, smells, sights, textures, and even tastes that you can call to the attention of your audience. In the early stages of making your Quest, a great way to conclude a group exploration of your route and/or box site is to brainstorm a list for each sense while the impressions are still fresh. You can then draw on this list when the time comes to write clues. Here's an example of a sound highlighted by the Lonesome Pine Quest:

> From the days of yore, an old town road you roam
> Look for a few fenceposts made of stone.
> The quacking and peeping you may hear,
> May be frog's music reaching your ear.
> Listen for the bubbly song sparrow's
> "Madge, Madge, please put on the teakettle" song
> Or the "wichity, wichity, wichity" of the common yellowthroat,
> You can hardly go wrong!

Smells are a rich part of any story, whether good or bad! Here's an example of a wonderful smell from the Fall Cider Quest:

Turn the crank, watch the apples crush!
And amber liquid flow.
Feel the earth, smell its rich full scent
And watch all the faces glow.

Use Humor

People learn best when they're having fun. Including local jokes in your Quest will draw people in and keep them engaged, as in the Vershire Village Quest below, which features the center of a very small rural town that is now being increasingly inhabited by ex-urbanites in search of the simple life:

The road that you're walking was once made of dirt
But now that it is paved, wear shoes and a shirt.

Here's another example of somewhat morbid humor. This clue is in a graveyard in the Cross Mills Public Library Quest:

You won't hear anyone complain so do
Look around closely for your next clue.

Weave in Literary Quotes or Quotes from Local Sources

Use direct quotes to make your Quest more authentic and to weave its story into a magical spell. As you saw in the Lyme Dances Quest, local research can easily make for a great story-based Quest. Here's an excerpt from Steve's original transcriptions of his interview with Ken Uline:

During the start of the war, they used to have card parties down to the church there. But the great thing was the Grange would put on a dance. Because, where were you gonna go? Gasoline was rationed; you had to have coupons to get gasoline, you had to have coupons to buy tires. Everything was tied right down. So here on Saturday nights we're having a dance in the church hall down here and in the Academy out there. One time the band was practically all my own relations: my brother-in-law played the fiddle, Charlie Balch played the piano, I played the banjo, my brother played the guitar and my sister called the changes. Whole families went to dances. Lots of times when they had the dance to Lyme Center, people down here would walk out to the dance and walk back home. Sometimes they had some rough times to the dances and sometimes they didn't. Some family was always bringing a jug of hard cider—that was one of the things they took up in the Second World War time.

And here's how it sounded translated into the Quest:

> As you walk down Acorn Hill, please pretend that there's snow,
> Though it's a dark winter night, there's somewhere special we'll go.
> It's the 1940s—and no, you are not traveling by car—
> But walking to the Academy where the dances are.

> *Ken Uline:*
> "One time the band was nearly all my relations—
> Myself, my brother, and brother-in-law; my sister calling changes.
> And whole families would turn out for the show . . .
> Sometimes with a jug of hard cider in tow!"

Demand Some Work of the Quester

Of course you want your finished Quests to be a generous offering to your community, but there's no law that says that you have to make it easy for your Questing audience. You might want to require a particular piece of research in the local library or city hall, for example, in order to complete the Quest. Or perhaps the Quester can only move forward with the help of directions from the local road commissioner, school principal, postmaster, or some other willing participant. Who would you like your Questing audience to meet? Which office or establishment would you like them to be aware of? You can use your Quest to expand their horizons.

End with a Bang

We have found that Questers have a palpable sense of excitement and accomplishment as they close in on the treasure box. It's worth taking pains to enhance this moment through your final clues. Here's an inspirational example from the Moose Mountain Quest:

> You have succeeded in your Valley Quest hike
> Armed with willpower, strength, and the like.
> Success is gratifying to all of us who
> Commit time and effort and the desire to DO!

10) Making Maps

WE LIKE MAPS! All maps. The maps of childhood: the end-page map of the "hundred acre wood" in A. A. Milne's *Winnie-the-Pooh*, the map of Wild Island in the *My Father's Dragon* trilogy. Antique, historic maps: topographical maps with the hills and mountains masquerading as flattened sea urchins, village maps with homes labeled "Riverside, Res. Of Capt. H. E. Brown," panoramic, hand-tinted maps from the nineteenth century. The word-rich maps that astute observers of nature make: Bernd Heinrich's map in *The Trees in My Forest*, David M. Carroll's map of Bear Pond in his *Swampwalker's Journal*.

When you make your Quest map, you just might be charting a small corner of the world—your special place—for the very first time. So before you begin drafting the map, we recommend that you consider four things:

> ➤ What is the primary *function* of your Quest map?
> ➤ What are the core *elements* or components—the things that must be included?
> ➤ Is there a *style* of map that is most relevant to your Quest site or theme?
> ➤ Is there a particular style or medium that brings you great joy?

As an example, for a Pinon and Juniper Quest traveling through forest on the Colorado plateau, the primary function of the Quest map might be to keep visitors from getting lost. Crucial core map elements come into focus in order to accomplish this. The Quest route or trail will be the central focus, with special attention given to side-trail forks and junctions. Prominent landmarks might be useful as well, to give visitors additional comfort through a general site orientation. Finally, images or symbols might highlight primary points of interest or danger. These points might also be numbered to cross-reference the map and clues.

In general, if the terrain of your Quest is more difficult—or out in the wild altogether—your map will focus more on the route itself. Greater use of landmarks will help your Questers know they are on the right track, and symbols, numbers, or letters help insure that Questers find all of your clues and comprehend the key points of your story.

The Schenectady Stockade Quest map, however, which moves visitors

Map of the pond

10.1 Map of Bear Pond by David M. Carroll. There are many ways to map a place. You can use words, images, or symbols; you can create a snapshot fixed in time or a record of memories generated over a lifetime. *This map is taken from Swampwalker's Journal: A Wetlands Year, by David M. Carroll, and is reproduced with permission from the author.*

through an urban landscape, serves a different function. Here, the concern isn't visitors getting lost; this map wants to convey a more general sense of the boundaries and elements of this historic neighborhood. In this Quest, the route itself may not be specified at all; rather, a matrix of blocks, buildings, and street names are its core elements, framing the area covered by the Quest.

PRIMARY MAP FUNCTIONS

All maps are two dimensional, graphic representations of space. Quest maps, however, perform a particular function: they work, in tandem with the clues, to steer the visitor from the beginning of the Quest to the treasure box hidden at the end. Some maps will be crucial to the solving of a Quest, while others are illustrations, or simply decorative. For a map to be a crucial element of a Quest, it must contain information that is not replicated in the clues. Those Quests that can only be solved by referring to both the clues *and* the map are the ones that are the most challenging and satisfying.

Illustrative maps might contain landmarks—drawings of the things you pass, for example—but this information is not vital to solving the Quest. The majority of the Quest maps made by school groups falls into this category. These maps are interesting to look at and convey rich content—they are good evidence of student learning—but they are not required to solve the puzzle. An intermediate or advanced Quester might be able to get to the treasure box without ever having to consult such a map. Decorative maps make your Quest more interesting visually but play no role in moving the Quester along the route.

You may choose any one of these approaches—crucial, illustrative, or decorative—for there is no best way to make a Quest map. Remember, too, as we discussed in chapter 5, that your map may focus on an entire Quest or a section of the Quest, or it may simply be a "treasure map" for the final clue. Just try and be clear about which strategy best serves your site, and which map function you have chosen. Then, working with that knowledge, execute your map neatly, beautifully, and with a sense of playful care.

CORE MAP ELEMENTS

As was noted in the examples above, your main map components will go hand-in-hand with your map function and in most cases reflect your Quest's theme as well. A city Quest will focus on the built landscape and include the relevant buildings, streets, stop signs, monuments, markers, manhole covers, and traffic lights, while a watershed Quest will portray tributaries, confluences, ridge lines, drainage basins, and points of safe water crossing. The former might include architectural details for orientation, while the latter consults landmarks like boulders, trees, bridges, or signs.

No matter where your Quest goes, you'll have to imagine a dotted line in space that separates your Quest site from the rest of the world and defines the realm of your Quest. The area *inside* that dotted line is the thing that needs to be portrayed by the background terrain of your map. Since everything that is there, however, cannot be included in your five by seven or eight by ten inch creation, you will have to hone in on the essentials. Using symbols instead of illustrations can help you fit more information into a smaller of limited space. These symbols or map elements become the foreground of the map.

MAP STYLES

You have many options for map styles: do you want to create your Quest freehand or utilize a technology like photography or GIS? You have a choice of materials: pencil, pen and ink, crayon, marker, watercolor, pastel. Finally, you have a wide array of artistic styles to utilize: representational, symbolic, expressionistic, humorous, or even cartoonish. A representational map might include the name of a bakery as a placeholder, while a symbolic map could include a sketch of a pie or baguette. A humorous map might use a fork, quite literally, to indicate a junction.

PERSPECTIVE

Even many adults can't make sense of the numerous lines that denote peaks and valleys on topographical maps! And while you might not have trouble with the "bird's eye" view of the shopping mall or airport, it's likely that your child, niece, or nephew has little idea what all those boxes, stars, dots, and lines mean. The map perspective—or point of view—you choose should take into consideration three things:

➤ Is there a perspective that best conveys this terrain?
➤ Is there a perspective you feel most comfortable drawing with?
➤ What do you think your viewers will be able to understand?

You don't need to spend too much time deliberating. A little forethought, however, can save minutes or hours at the drafting table.

David Sobel's book *Mapmaking with Children* is an excellent resource for

10.2 The Dana House Quest map, in West Lebanon, New Hampshire, is a slight elevation map. The map creates the feeling that you are looking at this scene from a second- or third-story window across the street. *Courtesy of Valley Quest: 89 Treasure Hunts in the Upper Valley.*

incorporating maps and mapmaking into a family or group Questmaking process. David's unique contribution to the field is his connection of children's developing spatial awareness with both the terrain they can conceptually explore and the style of maps these children can create and comprehend. David notes that five-year-old children tend to make pictorial maps and can comprehend the landscape of their *homes*. By the age of seven, children progress to slight elevation maps that can include the broader geography of *neighborhood*. Around the age of nine, children are able to move on to mapmaking strategies that include an elevated viewpoint such as high oblique, as their awareness expands to the larger conceptual frameworks of *community* and/or watershed. Most eleven-year-old children have developed to the point that they can visualize their neighborhood within the context of their city, their city within a state, their state within a country, and so on. Versed in the conceptual framework of "nesting boxes," they can create and follow accurate aerial or bird's-eye-view maps.

Your Quest map and map style will depend on the age of the group you are working with, the characteristics of the physical environment your Quest is exploring, and the age of the audience you are trying to speak to.

BOULDER
WOODS
STONE
FOUNDATION
SPLIT
BIRCH
N
S
E
W
MILL
SITE
BROOK
MILL

10.3 The Blacksmith Bridge Quest map is a high oblique map. The viewer floats up at a high angle, looking down upon the scene. The Quest site nests within the larger frame of the forest. *Courtesy of Valley Quest: 89 Treasure Hunts in the Upper Valley.*

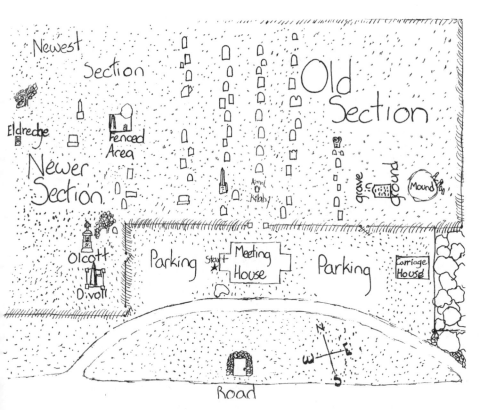

10.4 The Rockingham Meeting House Quest map offers a bird's-eye view of the site.

Courtesy of Valley Quest: 89 Treasure Hunts in the Upper Valley.

AUXILIARY MAP ELEMENTS

Regardless of the map style you choose to adopt, your Quest map should include a compass rose indicating north (to help orient visitors with directional clues) and a key to symbols you used on your map. Your map also may include a scale and contour or elevation lines, if appropriate. You can make your Quest map more attractive by adding a decorative border.

The Compass Rose

Every Quest meanders through a place, and at least half the time your clues will be passing on directions, or "movement clues," to your visitors. These directions might be based on body orientation ("at the light, please turn right")

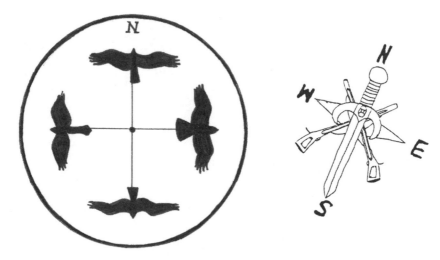

10.5 a and b The Gile Mountain Quest compass rose drawn by Ginger Wallis references the silhouettes of four types of hawks likely to be seen from Gile Mountain's fire tower. The Woodstock CWM Quest compass rose, drawn by fourth-grade students at Woodstock Elementary School, offers subtle clues that hint at the meaning of CWM. *Courtesy of Valley Quest: 89 Treasure Hunts in the Upper Valley.*

or they might be physically based ("follow the brook downstream"). It is reasonable to assume that any group of visitors will notice the traffic light or be able to tell their left from their right hand. However, to increase the degree of difficulty (and fun) of your Quest, you can add a compass rose indicating the cardinal directions. Adding a compass rose allows you to create directional clues ("heading west I think is the best"). The compass rose is a key link between Quest map and clues.

Your compass rose may be as simple as a cross or an arrow; however, our favorite compass roses reflect some inherent quality of the site or the story. The Gile Mountain Hawk and Haiku Quest leads up to a tower that birders climb to watch the hawk migration throughout September. The compass rose for this Quest references the silhouettes of the four types of hawks likely to be seen soaring overhead: falcon, buteo, accipiter, and eagle. The compass rose can be straightforward—an element lifted or repeated from the map—or it can be tricky. The Woodstock CWM Quest compass rose offers subtle hints that CWM is going to mean "Civil War Memorial," but not so much information that the initials give the treasure box location away.

In creating your compass rose, the possibilities are endless. But the main point is that the compass rose offers you the opportunity for creativity (in coming up with a design concept) and artistry (in executing that design) while deepening visitors' connections to the qualities of your chosen Quest site and theme. Layers of symbolism and pattern allow a certain sense of depth to arise. Christopher Alexander writes in *The Timeless Way of Building* that if things "get their character from the patterns they are made of, then somehow the greater sense of life which fills one place, and which is missing from another, must be created by these patterns too."[1] Repeating patterns support recognition and learning and are key factors in creating Quests that come alive.

Symbols and Map Key

In order to reduce the overall size of the map, the mapmaker must use symbols to represent the things themselves. Some symbols are obvious enough that they need no interpretation. There will be maps, though, that cover terrain complex enough to require the development of symbols and a key. Use of the map key is essential to solving the Lyme Sheep Quest and arriving at the treasure box in the cellar hole of the old sheep farm on Cole Hill. This Quest travels through deep woods, and the symbols for natural and cultural features—fallen birch trees, stone walls, and apple trees—help prevent the Quester from getting lost.

Borders

During the course of making your Quest, you will make a lot of observations. You may also generate a number of detailed sketches of some of the more distinctive attributes of your Quest site. Placing a border around your Quest map—or text—puts these artistic elements to use. The border may be purely decorative, or it can serve as a checklist of features: architectural motifs, the numbers you will see, types of insects, or the tracks of animals common in the habitat the Quest explores, for example.

MAKING MAPS WITH A GROUP

If you are making a Quest by yourself or with a small group, mapping your Quest won't present a problem. However, if you are working with a large group and wish to have every participant contribute to the mapmaking process, we suggest you break your map into sections and have the small groups adopt and create different map elements.

In Piermont, New Hampshire, students adopted a number of historic buildings in the neighborhood adjacent to their school. The first step of the map-

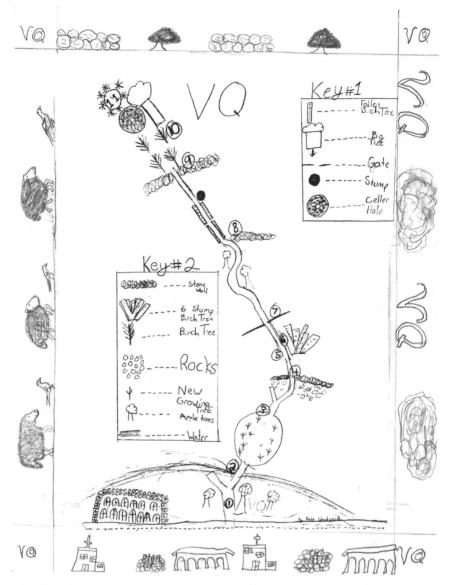

10.6 The Lyme Sheep Quest key, drawn by fourth-grade students in Lyme, New Hampshire, uses symbols and milestones to steer visitors to the stone foundation of an abandoned sheep farm on Cole Hill. *Courtesy of Valley Quest: 89 Treasure Hunts in the Upper Valley.*

Quest Map

Follow the rock wall that is plain to see up to small wooden small stairs. Pay no attention to our warning just follow without any cares.

Up the stairs and hang a right, follow the path 'till the field's out of sight.

The yellow dot path is the next clue. There aren't any purple, green, red, or blue.

Keep on your path up the hill to patch that hemlock does fill.

Through the Hemlocks you will see a tiny clearing just made for you and me. Head for the clearing and walk on through past the "W" and the old hollowed out tree.

Continue straight ahead if you please, but be on the lookout for a "V" that stands proud and tall.

As you stand here at "V" face 320 degrees take 50 paces or 100 large steps passing the old maple trees.

Make a left through the stone entrance toward the white pine. Placed on top of the rocks you will see the cups from which pixies drink wine.

Now stand at their cups and look for white quartz rocks.

Once you find the white quartz rocks you will find the treasure box.

Hemlock

VINS

Start Here

Pond

10.7 Decorative borders enhance the visual appeal of your Quest and may hint at the theme.

ping process was to meander through the village as a group, filling in a worksheet to determine what the group already knew about each structure. This process served three key purposes: general site orientation, preliminary research, and a first opportunity for site bonding. Developing a simple worksheet with blank squares to draw in and leading questions to answer helped to keep students focused on the task at hand.

By the second visit, groups were ready to choose or adopt the building that

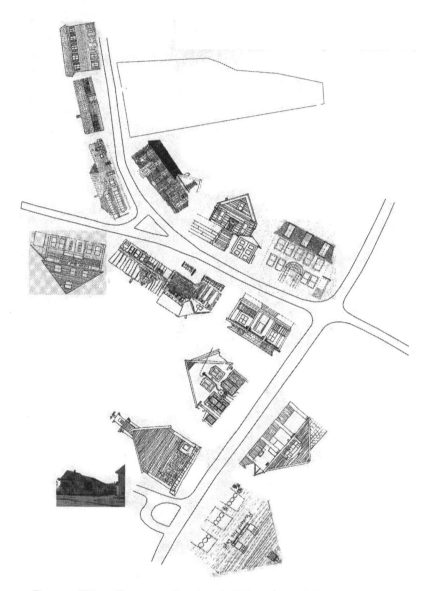

10.8 Piermont Village Quest map. Drawings by Piermont third, fourth, seventh and eighth graders. Students plotted GPS points in Piermont, New Hampshire, to create a computer-generated map featuring scanned and downsized reductions of their drawings. *Map © 2002 by Donald Cooke, Geographic Data Technology Inc. (GDT).*

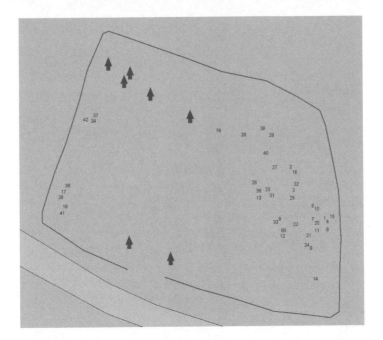

10.9 Beal Cemetery Quest map. Points plotted by Lyme fourth graders. A Trimble unit, loaned by a business partner, produced a high-quality map of the cemetery. Each point on the map represents an individual tombstone location. *Map © 2002 by Donald Cooke, Geographic Data Technology Inc. (GDT).*

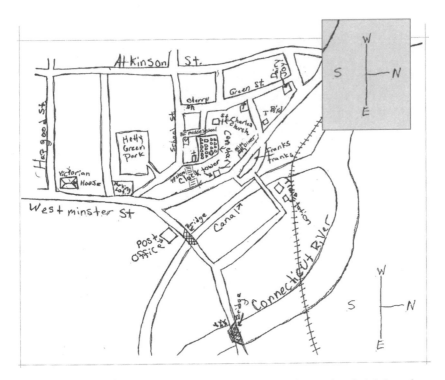

10.10 Bellows Falls History Quest map by Compass School seventh and eighth graders. This map was created by referencing a pre-existing map of the downtown. Key points of interest were added through captions and arrows.

10.11 Haverhill Corner Quest map. Enlisting the participation of a local artist can lead to a beautiful Quest map. © 2002 by *Allianora Rosse.*

10.12 a and b George Stowell Library. Drawing by Clara Lipfert. Photograph by Simon Brooks. Maps can be built up from individual drawings. The accompanying clue reads "George Stowell gave money for this place of books / Right next door is the place for crooks!"

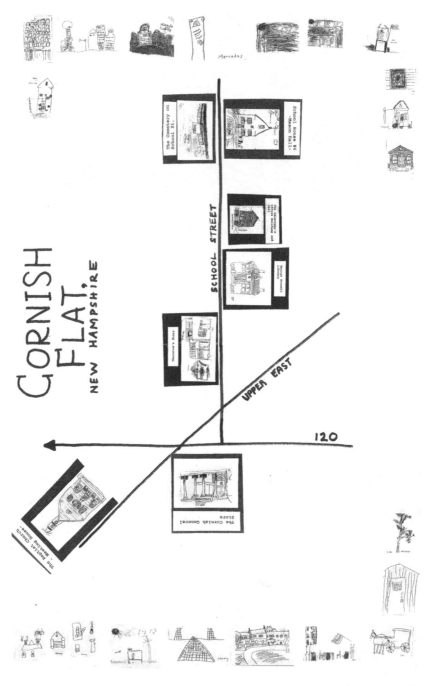

10.13 Cornish Flat Village Quest map by Cornish, New Hampshire, third graders. Seven student drawings have been combined to make a map.

most appealed to them. The groups then went back out in the field to make sketches. In addition, each group used a digital camera to take a photograph of their building. The photographs allowed students to bring images back into the classroom and refine their drawings over time, deepening the quality of both the artwork and the resulting Quest map.

Global Positioning System (GPS)

Each group also used a GPS unit to record the precise location of their building on the surface of the Earth. Global positioning systems record the position of a small, handheld unit on the surface of the Earth by "triangulating" that positions in relationship to the positions of satellites orbiting the Earth. Higher quality (and more expensive) GPS units depend on more satellites, while inexpensive models depend on fewer. As the number of satellites referenced increases, the accuracy of the GPS unit increases. Regardless of quality, GPS units are excellent tools for group mapmaking and can help group members increase their understanding of both mathematics and technology while working on their Quest map.

Some Other Group Mapmaking Strategies

If your Quest is set in a village, town, or city center, chances are good that a map of this terrain already exists and is readily available. After locating such a map, you can focus in on your chosen neighborhood, enlarge that section using a copy machine, and then trace or draw in the core, compositional elements.

The intimacy fostered between a group and a place through the creation of a map is conveyed by the Cornish Flat Village Quest, created by third-grade students in Cornish, New Hampshire. In this Quest, students worked in groups of two to four. Each group adopted, studied, wrote clues about, and illustrated one important location in the small village. The seven sites were then joined together by the map.

There are many kinds of Quests—some will require no map, while others will need detailed maps. Still other Quests will simply require illustrations that accompany the clues and offer visitors enough concrete visual information to steer them along the true path. The map you make should be accurate and beautiful, but the rest is up to you.

11) Treasure Boxes, Stamps, and Sign-In Books

IT'S HOT—NINETY-TWO DEGREES IN THE SHADE. You've been out pounding the pavement for an hour (well, that includes stopping briefly for a root beer at the pharmacy), but excitement is running high. You're closing in on the treasure box—you're sure of it.

"Let's see, it says:

> "Take your answer for 'A' and add it to 'B,'
> Together they'll give you an answer that's 'D.'
> Now subtract 'D' from 'C,' but don't tell a soul,
> Go to this number on your map, you'll be at your goal."

"Ummmm, I get eight. How about you?"

"Yeah—the same. Let's see . . . there's eight on the map. Down in back there. Let's go!"

You head down the cracked cement stairs past the town swimming pool and then along a narrow strip of grass between the chain link fence and a stone wall. The Ottauquechee River is easing over rocks below. The strip of grass narrows, looking more and more like it doesn't lead anywhere.

"Can this be right? Are you sure we're allowed back here?"

"This is exactly where eight is on the map. Come on, it should be here."

"But this path ends at a fire escape. It wouldn't be hidden on a fire escape—that's crazy."

"Wait! Look at this little door in the stone wall . . . this must be it!"

"YES! We've found it!"

You open the tiny wooden door with its porcelain knob and find a plastic refrigerator box inside, bearing a label that states "This is a Valley Quest Check-in Box." Inside is a homemade book covered with flower print fabric, a stamp pad, a pen, a handful of colored pencils, a pencil sharpener, and a homemade rubber stamp.

"Check out this stamp. It looks like Ratty from *The Wind in the Willows*—see his ears and his little rowboat?"

"Yeah, and here's an inscription in the book:

"Welcome to our Valley Quest box,
Today we hid it with cement and rocks.
It's sunny and warm and the river sounds sweet.
We checked it out with our bare feet.
    —Signed, the River Rats"

"It says the box was first put out in 1997. This is their third sign-in book!"
You find a little laminated essay taped to the inside of the box cover.
"Did you see this sheet? It says this building used to be a woolen mill but
now it's a theater. It was built in 1796. And that building has a one-lane bowl-
ing alley in it. It says there's an info rack inside the building, too, if we want to
know more."
"It's so peaceful here by the river."
"Look, here are some colored pencils. I think I'll make a little sketch in the
sign-in book. Do you have our stamps so we can stamp in?"
"Yup, I always keep them in the side pocket of my backpack here, just
in case."
"Hey, see how great the Ratty stamp looks in my passport? Now I only
need three more and I'll have twenty and get the patch."
"This is so cool!"
Questing is low tech and low cost. It requires only a minimal amount of
equipment, most of which you can make yourself. In this simplicity lies the potential for beauty, and for streamlined functionality. Let's take a look at the primary tools of the Questing trade: treasure boxes, stamps, and sign-in books.
We'll also take a look at personal stamps and passports and a few other things
you might pack into your treasure box to make it even more fun.

THE TREASURE BOX

Nothing beats finding a treasure, but a wet, stinky, moldy box filled with runny
ink and a fuzzy green stamp pad hardly qualifies as a treasure. You obviously
want a treasure box that is not only inviting and filled with interesting lore
and supplies but also functional and easy for repeated visitors to use and re-
place. It needs to fit the shape of the hiding spot: for example, you'll need
something short and round to fit inside a pot of fake flowers in a cemetery, but
something fairly flat to fit behind a hinged sign on a brick wall. You also need
to consider what, if anything, you're hoping to fit inside the box along with the
basic logbook, pencil, stamp, and stamp pad.

If you've got an outdoor location, the trick is to find a treasure "box" that will not only work well in pleasant weather but hold up to the worst your region has to dish out. In northern New England, that means rain, snow, humid heat, and ten below zero, all within our April through November Questing season!

Plastic Refrigerator Boxes

Standard flat, rectangular plastic freezer containers are probably the items most frequently utilized as Quest treasure boxes. They are affordable and readily available in any supermarket, and they work well enough with the exception of an occasional crack or leak. If you live in a place where the weather is dry and the temperature fluctuations are moderate, you can have good success with them. Take care to be sure that they hold a seal. If you live in a wet or cold area and choose to go with one of these boxes, we suggest putting your logbook and stamp in a resealable sandwich bag inside the box, and then put the whole box inside a larger box or gallon-size resealable plastic freezer bag. Be sure to include a reminder in your logbook to carefully reseal the box before stowing it away again.

Lexan Utility Boxes

If you can afford them, waterproof and unbreakable Lexan boxes make a great upgrade. One of their advantages is that Questers are more likely to close them tightly. They also have more room for activities, resources, and supplies, though they are not quite flat enough for some hiding spots. These boxes are available from camping outfitters or recreational equipment stores.

Plastic Screw Top Bottles

For tall, narrow hiding spots, screw-top plastic water bottles make a fine alternative, especially if you get the sturdy kind designed for hiking. Test your sign-in book to be sure it rolls and can be slipped in and out of the bottle. You'll also need a smaller inkpad, or use a felt-tipped marker to color the stamp.

Recycled Containers

In situations that are protected from the weather, recycled household containers may work. These have the advantage of modeling an ethic of reuse and will certainly save you money. We've found that a one-pound coffee can fits well in

11.1 Quest boxes need to be weatherproof, large enough to hold what you want to put in them, and small enough to hide. There are many options; here are a few. Photograph by Simon Brooks.

many bluebird houses. The plastic lids need to be replaced on a regular basis, however, as they tend to crack. On Dartmoor, ice cream containers are sometimes used as letterboxes, and these work well enough if you insert a plastic bag liner, though these might only last a few weeks. Whatever you do, just be sure that your treasure box won't be mistaken for trash!

Indoor Options

Indoors, you can let your imagination go wild. We've hollowed out the inside of an old book for use in a library, hidden a small oak chest behind the concierge's desk in a historic inn, and loaded up a cigar box in a general store. We've heard about someone who has plans for a pair of faux paper towel dispensers in a restaurant's restrooms—with different stamps for boy and girl Questers!

Whichever kind of box you choose, you will want to supply it with a label on the outside that identifies it as a Quest treasure box and gives directions and some contact information. If you have an email or website address, you might want to include those as well. Your choice and design of the contents of your

This is a Green Trail Quest check-in box.
Please do not remove or disturb it.
Please make sure that you leave it hidden
for others doing the Green Trail Quests.
Thank you!!

For more information on Green Trail Quests,
Contact:
Green Trail Quest Central: 401-364-3878

11.2 Green Trail Quests label. Remember to label your treasure boxes and be sure to include a contact phone number or email. If you have a website, include that as well.

box give you lots of room to express the special features of the place. Let's look at each of these contents separately.

STAMPS

Of course, the "treasure" in Questing is really the walk, the discovery, the learning, and the fun, but in the end it does all come down to the collection of an impression of a stamp, so you want your stamp to be a good one. The general idea is that the stamp is an icon representing the essence of your place. As with any symbol, sometimes the simplest idea takes the most thought, but if you pull it off, you will have a sought-after box with a lot of personality that is well worth the hunt. The best stamps are either homemade or personally designed and commissioned from a rubber stamp artist. If you're short on time or creativity, or if you're lucky, you might also find the perfect image commercially available at a local rubber stamp shop or on the web.

Many people find that they enjoy the stamp element so much they change

11.3 Vermont History Quest stamps by Simon Brooks and Steve Glazer. Four stamps—hidden in four boxes—fit together to create a single composition.

the stamp in their Quest box every year, thereby encouraging repeat Questers. Others keep one element constant, for example, a border, while shifting the internal elements seasonally. We've seen wonderful multipart stamps that have a border, a pool, and a fairy that can be combined in different positions and printed in different colors. (You can equip your box with a multicolored stamp pad.) We have also made a four-part stamp and hidden each part in a different box. Questers were required to collect all four to complete the image.

11.4 Whoops! Guess who forgot to reverse the letters on their stamp? Remember that your stamp will print a reverse image; remember to consider this as you plan your design. Stamp and photograph by Simon Brooks.

If you choose to make your own stamp, remember two key facts and you'll do fine:

➤ The stamp will print in a mirror image, so every part of your composition, especially the letters, must be reversed.
➤ Your design can be a positive or a negative image. Decide before you make your stamp whether you'd like to print, for example, an ink turtle or a white (blank) turtle framed by an ink outline or frame.

We've outlined a range of techniques that work well. Your choice will depend on your comfort level with sharp carving tools, the amount of time you have, and your interest in collecting and utilizing recycled materials. Your choice of medium may dictate the size of your final stamp, but in any case, be sure to make it no larger than the size of a conventional stamp pad: two by three inches (four by six centimeters).

Start by drawing a series of boxes on a sheet of blank paper. Make the squares approximately the same size as your stamp material. Play with some ideas by sketching them in different boxes until you come up with a concept

11.5 When you cut your stamp, you also have a choice between creating a positive or a negative image. Stamps by Delia Clark.

you really like. You might want to get your mind focused on iconography by first flipping through your field journal, source book, or magazine. We recommend that you keep your design very simple at first: a leaf, the sun, some initials, maybe a border. Choose your favorite design and refine and stylize it as necessary, then create the stamp using one of the techniques below.

Carved Rubber

You can make a very crisp, detailed, and professional-looking stamp by carving vinyl erasers or an art medium called "carving block." Carving block's advantage is that you can cut and size your material to suit your composition. Erasers have the advantage of being easily available in your local office supply store. Be sure to select a stiff white vinyl eraser such as Magic Rub, PZ Kut, or Staedtler Mars, not one of the old-fashioned pink or brown crumbly ones. Most people prefer to carve using V-shaped or U-shaped gouges designed for linoleum, or straight-edged carving knives.

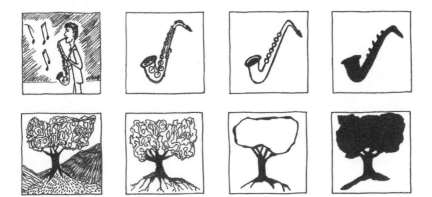

11.6 First try out a number of ideas. Then, pick the one you like best and refine that image. Drawing by Kal Traver.

11.7 a, b, c Rubber stamps can be quite beautiful. Try to reference a quality of your Quest site in your stamp design. Stamps designed and cut by Simon Brooks.

Start by transferring your design to the eraser. The easiest technique is to transfer the image onto tracing paper with a heavy leaded pencil, then center the design over your eraser and rub hard on each of the lines. They'll show up crisply on the eraser. If you don't have tracing paper, you can accomplish the same thing by drawing your design heavily on regular paper, moistening the eraser (saliva, by the way, works fine), and pressing the two together. You can also start with a photocopied image. Trim the design and center it, image side down, on the eraser. Moisten a cotton ball with acetone nail polish remover and gently rub the back of the page.

Carve around the design, taking away everything that's white and leaving everything that's black (or vice versa if you want on outline design). Begin with the fine, detailed areas first, carving away from the design rather than into it. If you're using a straight-edged blade, hold it at an angle away from the design so that you don't undercut the edge. Go slowly, as it's a reductive process, like cutting hair. Once you've cut something off, you can't put it back on! You can accomplish shading by making tiny vertical cross-hatching cuts, which will show up when you print.

As you carve, test your image repeatedly by stamping it on a colored stamp

11.8 Simple stamps may be cut from adhesive felt or Dr. Scholl's adhesive pads.

pad and seeing how it looks on paper. The color will remain on your eraser and show you where you still need to carve.

Adhesive Felt

Though not as detailed, Dr. Scholl's of type peel-and-stick footcare foam makes serviceable stamps quickly and is especially well suited for use with small children. Self-adhesive soft rubber material is available from art supply stores. Either medium can be easily mounted onto a small wooden block. Follow these instructions:

1. Trace or draw your design onto the back of the material. No need to reverse it, as it will print reversed.
2. Cut out the shape using a pair of scissors. If you are going for an outline, begin by snipping the folded center part so you keep the outline intact.
3. Sand your small wooden block smooth, if need be. (This is a great way to keep early finishers occupied.)
4. Peel and stick the design onto the block.

Reused Materials

If you want to focus on reusing found materials, try looking for some old inner tube in your garage, on the side of the road, or in a bike repair shop. You can cut it into quite detailed shapes and glue it onto wooden blocks using household rubber cement. Another, less permanent option is to work with styrofoam packaging. In addition to cutting out shapes, you can get some interior detail

11.9 a, b, c, and d Stamps can also be made from a variety or recycled of found materials. Here you see, left to right, stamps made from a potato, a styrofoam meat tray, a finger print, and a bicycle inner tube.

by pressing in lines and dots with a pencil or fork. If you really get into it, you'll find lots of other materials, including corks and balsa wood, that stamp well alone or as part of a multimedium stamp. Just remember, you'll need to monitor your box more frequently if your stamp is fragile.

Temporary Stamps

If you just need a stamp for a brief time, try slicing a potato or yam in half, then carve the flat side. Carving a travel-sized bar of soft soap is a possibility, as is printing with your own thumb and adding a few appropriate decorations afterward.

SIGN-IN BOOKS

There's something very satisfying about telling the world "I was here," and the sign-in book is your way of letting people do this. It's also a way to welcome people to your site and let them feel a sense of camaraderie with those who've been there before them. These quotes from the Boundary and Fox Ponds Quest sign-in book indicate just that:

Pretty day, never resting. Looking, laughing, always questing.

Although I'm 71 years old, I enjoyed this Quest just as much as you young folk.

11.10 Signing in. Photograph by Jon Gilbert Fox. Sign-in books play an important role in a Quest program. Here, visitors can share their experiences and stories.

We've found that unlined paper works best, as it encourages sketching and also shows off Questers' personal stamps to advantage. You can find small, un-lined sketchpads in art supply shops or, if you have a bigger box, you can use an unlined blank book. If you do a good enough job with your box, you shouldn't need to use waterproof paper pads, which have a tendency to smear the stamp impressions.

We also encourage you to consider making a sign-in book yourself. Homemade books are attractive and durable and give visitors a sense of being personally welcomed by those who put out the box. There are lots of ways to make a book—here's one:

1. Collect the following materials:

    Boxboard, such as an empty cereal box, cut into two pieces, each $4\frac{3}{4}$ x 6 in. (12 x 15 cm)

    Ten or so sheets of standard copy paper

    One sheet of colored construction paper or colored bond

    One piece of fabric 8 x 11 in. (20 x 28 cm). You might want to choose a pattern that matches the theme of your site, or make your own pattern using fabric paints or fabric crayons

    Glue, tape, and scissors

    Needle and thread or long-armed stapler

2. Cut the copy paper in half widthwise and fold in half again, forming the pages of a small book measuring $4\frac{1}{4}$ x $5\frac{1}{2}$ in. (11 x 14 cm).
3. Cut the construction paper in half and fold one half to form a cover around the pages—this will become the book's end papers. Reinforce it by running a strip of tape along the outside of the fold.
4. Tie a knot in a length of thread, leaving a tail, and thread a needle. Starting on the back side of the book, poke the needle through all of the thicknesses of paper on the fold. Continue to stitch along the fold, taking large 1-in. (2-cm) stitches. When you come close to the end, reverse your way back to the starting point, using the same holes. Tie the ends of the thread together using that long tail you left. The knot should be on the outside of the fold.

    If you want a faster technique, use a long-armed stapler and staple along the fold!
5. Lay out the fabric wrong-side up and center the boxboard pieces side by side on the fabric with about $\frac{1}{2}$ in. (1 cm) between them. If you're using a cereal-type box, it's best if the print side is up so it won't show through the fabric.
6. Glue the boxboard in place. Fold down triangles on the four corners and glue them in place. Fold in all four edges and glue them securely in place. Make sure to glue the fabric area near the corners.

7. Nest the pages inside the cover and glue the end sheets to the cover, leaving some loose fabric along the spine to allow for easy opening.

SEIZING THE TEACHABLE MOMENT: OTHER BOX CONTENTS

You can leverage the excitement Questers feel at finding your box into a great learning moment for them by including additional information, material, or challenges in the box. In fact, it was this idea of challenging box finders to do an activity once they got there that started the Letterboxing tradition of writing poetic entries in sign-in books. As Anne Swinscow explains it in *Dartmoor Letterboxes*, "Some years ago a box was put out where everyone who signed in was asked to write a poem in the book. Since that time poems and letterboxing seem to have gone hand in hand, and letterboxers break into verse at the slightest provocation."

Want to Know More?

At the very least, we suggest writing a little essay that tells "the rest of the story" or a list of amazing site facts. You can laminate this sheet and glue it onto the underside of your box cover, where it demands to be read. If you're more ambitious, you can add quotes, photographs, diagrams, and other material that references the things they've just seen along their route. You can also leave a stack of brochures about your partner organizations in the box, with an invitation for visitors to take one if they are interested.

Questions and Challenges

You can lure people into studying the material in your box by posting a list of questions with your Quest and supplying the answers in the box or in a series of boxes, as we did with Valley Quest:

> What connection does the famous children's book author Dr. Seuss (Theodor Geisel) have to the Upper Valley?

> The Upper Valley was home to the country's largest community of Shakers during the nineteenth century. What cardinal principal did they base their lives on?

A recent survey selected an Upper Valley town as the second-best small town in America. Which town?

Somewhere in the Upper Valley roost three big rocks known as the "Big Hen and Little Chicks." Where are they?

Another good approach is to include a further challenge to Questers once they find the box. South Shore Quests included the Burbank Boulder Challenge in a treasure box:

How would you go about finding the volume of this rock, in cubic feet? Hint: If you measure the height, width, and length and multiply them, you can get the volume (roughly) in cubic feet.

How much does Burbank boulder weigh? Hint: A cubic foot of water weighs 2.18 pounds, and a cubic foot of rock weighs 7.78 pounds. A ton is 2000 pounds.

Our challenge: Leave your estimate of the weight of the boulder in the logbook in the Quest treasure box with your name, age, and address. The most accurate figure at the end of the year will be recognized in our newsletter, and we will send the winner a prize.

True to their promise, the next winter's newsletter announced the prize: a $10 gift certificate to their local rubber stamp store!

Supplies

Encourage closer observation of the local environment by including inexpensive hand lenses, bug collection boxes, field guides, and other study equipment in your box. If the box is in a relatively secure location, you might even include small binoculars. Questing and ornithology make great companion activities on a walk through a natural area.

> Can you hear the "laugh" of a robin?
> Or the "wrock" of a raven?
> Both of these birds
> Inhabit this haven.
> —From The Lonesome Pine Quest

You might even want to make a bird stamp to match this theme. You can encourage creative logbook entries by supplying colored pencils, a small tray of watercolor paints, or crayons.

More Quests

Include a promotional flyer in the box inviting visitors to do other nearby Quests, or a tantalizing advertisement for a Quest that will be released soon. You can keep them Questing near your site through adding spur Quests (described in chapter 5). You might also find, through no doing of your own, that a floater Quest has appeared in your box . . . just passing through.

Prizes

Though we generally steer away from creating the expectation that Questing involves any more reward than the joy of the experience and the fun of collecting a stamp, we have found that occasionally a small prize makes sense. Some people have enjoyed placing a gift such as a tiny jar of honey, a charm, or some polished stones in the box for the first visitor of the Quest season. We've also used strategically placed and well-advertised caches of gold dollar coins to begin the Questing season.

EQUIPMENT FOR THE QUESTER: PERSONAL STAMPS AND PASSPORTS

In New England there is a firm tradition of "peak-bagging" or climbing all of the mountains over 4,000 feet in a given range. Similarly, we have the 251 Club, people who have visited and collected a photograph from every incorporated town in Vermont. We've found that this urge to "collect them all" applies to Questing as well. We suggest that prospective Questers make or buy themselves a personal passport so they can collect all of their stamp impressions in one place. The passport can become a journal to keep track of field notes and drawings from the Quest experience as well. You can guide people in making a thinner version of the sign-in book described above, or print up some bordered stamp-collection sheets with your Quest program information at the top.

Once they get started, Questers usually acquire or make a personal stamp to use each time they sign into a Quest box. This is an artful way of personal-

11.11 a, b, c Questers like to create stamps of their own, which they use to personalize their sign-in book entries.

izing their signatures, and sometimes it even takes the place of their real name. Those who get into it often just sign in with their stamp and a "handle" or nickname to go with it, such as "Golden Mask," "Count Olaf," or "Shining Brow." Regulars come to recognize each other's stamp names and seem to get quite a kick out of chance meetings with the people behind the names.

Offering stamp and bookmaking workshops based on the techniques described above is a good way to build public participation in your Quest program and get people started on what may quickly become their latest obsession.

12) Sharing Your Quest

THE QUESTING PROCESS YIELDS two treasures: the initial treasure of your discovery and learning and the secondary treasure that comes from sharing. It is wonderful—and enough—to really explore and get to know a place by and for yourself, for in the midst of that inquiry-process emerge experiences, insights, and perhaps even feelings of deep appreciation or love. But the sharing of that treasure allows the circle of good feeling to radiate out, rippling across the community and affecting the lives of others. This is a precious gift.

> October 8, 2001
>
> With great thanks for this day
> and for the chance to share
> this gorgeous place with
> our special friends.
>
> Lindsey and David
> —A note left in the Gile Mountain Quest sign-in book

If you plan to share your Quest, you will need to come to a precise conclusion with your map and clues, bringing all of your work to fruition. If your map is off or your clues incomplete or inaccurate, visiting Questers may have the experience of getting lost rather than of discovering something. If your clues are strong, the map is strong, and the treasure box is placed securely out in the field, however, you have created the opportune conditions for community discovery, fun, and connection.

> November 23, 2001
>
> Gorgeous views,
> Wonderful hike &
> Delicious Apples.
> We'll be back!
> —A note left in the Lyme Pinnacle Quest sign-in book

Far more than half of the benefit of this program comes from the sharing. There are few things more enjoyable than giving the gift of knowledge or experience—the sharing of what is special to you. A series of Quests exploring your place becomes a true, vital community asset, in the same way that a park or museum serves as a community asset. A Quest program can become almost like a community library: a collection of habitats, historical sites, neighborhood walks, and thematic, pedestrian adventures focusing on the important stories of the community. Once public, the Quest is a new place to visit and learn about. It's a place to bring visitors, or an activity to do with a scout troop. Out on the Quest, many connections are made. There are intimate, timeless, unplanned experiences—the glint of summer sun off the reservoir, or the smell of rain in the ferns at the city park, the feeling of a spotted salamander in the palm of your hand, or the taste of sticky baklava fresh from the Greek bakery. Questing generates memories and the feeling of being at home. This is a warm feeling that lasts.

8/31/02

It was an overcast day
& we watched a grey heron
(though probably named blue)
As it flew away
Across the mud where cattails grew.
We, a child and a parent,
Our one-to-one morning almost through
Felt magic in the dry grass,
The golden rod, the Queen Ann's lace;
After many days have come and passed
We will remember this place.

Patricia and Seth
—Entry in The Lonesome Pine sign-in book

Walking together out on the land is an ancient human practice. Bruce Chatwin writes extensively about this venerable tradition in *The Songlines*. Walkabout was something we did, as a species, for millennia. We walked in places, sang in places, read our way, using natural and cultural features, through the landscape and through our lives. Our existence depended on these stories and songs: here was where you could find water; over there, food. Culture was oral and the landscape was written in stories.

Perhaps we don't do this anymore—sing our way down Main Street—but still, inside, is there perhaps a memory of this? Or a longing for it? The public Quest creates a structure within which people can be out on the land together, wandering, curious, talking, learning, connecting with the environment in a meaningful way. And then comes the great remembering: the recollection of the bigger story lurking out there just beyond the privacy of our interior lives, the bigger story including human history, geologic time, seasonal cycles, and migrations, or what Judith and Herbert Kohl in *The View from the Oak* call the "private worlds of other creatures."

STRUCTURES FOR SUCCESS

In order to successfully share your Quest, you will need to take care of a few important things: permissions, testing, monitoring, publication, and distribution. (Permissions are discussed in chapter 4.)

Testing Your Quest

Your Quest won't work as a public document, or generate community good feeling, if it does not work as a Quest. While everyone loves a Quest, few people enjoy getting lost, and most people get frustrated if they can't make sense of a clue. In order to be certain that it does succeed—that the clues and the map perform their desired functions—test your Quest with several people who were not involved in the creation process. Better still, find someone who is unfamiliar with your site and story to test your Quest.

When your volunteers test the Quest, don't give the clues and directions away to them. Giving hints is easy to do, with only a glance or a turn of your body. Follow a little way behind your Quest-testers and observe when and where they have confusion. Make notes so that you can improve your clues or simplify your map. During the Quest testing sessions, be sure to record the length of time it takes to complete the Quest, as this is something you'll need when it comes time to publish. You can also ask your volunteer to complete a small check-box survey, recording the time they spent on it and answering the following four questions:

1. Rate the physical difficulty of this Quest: easy, moderate, or difficult
2. Rate the mental challenge of this Quest: easy, moderate, or difficult
3. I would recommend that visitors to this Quest site bring the following things:
4. Suggestions for improvement.

Box Monitoring

It is disappointing for visitors when they are certain that they are in the right place, and the treasure box is not there. The success of a Questing program depends on loyal volunteer box monitors.

The job of a box monitor is to check on the treasure box every four to six weeks (or more often) during the Questing season in order to make sure that the box is in place and complete. Sometimes pens or stamps disappear, ink pads run dry, or sign-in books fill up. Animals can (and have) discovered, opened up, and/or gnawed through the contents of the treasure box. A regular visit by the box monitor keeps the box well tended, complete, and welcoming for the visitor.

The box monitor can develop a creative, personal relationship with their adopted Quest treasure box. Boxes might be improved and personalized by the addition of art supplies, field guides, laminated photographs, copies of a favorite poem, or lists of recommended activities. One box we know shows visitors how to make a found-object sculpture and then invites them to make one. Another Quest ends up in "cairnville": at this site, everyone builds interesting rock sculptures.

Some Quest treasure boxes focus on animal observation. Here is an entry added into a sign-in book by a box monitor on his regular box check-in.

4/19/02
74 degrees

1 Canada goose
3 hooded mergansers
2 song sparrows—singing
1 swamp sparrow—out in the marshes
1 white-throated sparrow
1 common raven—"wrock, wrock"
2 belted kingfishers—chattering
1 pileated woodpecker
2 tree swallows

—BS

The box monitor, in visiting the Quest site again and again, becomes the person who really knows this particular place. Their personality and dedication can lend each individual Quest box and sign-in book a particular character or feel. Box monitoring is itself a path to intimacy with the land and stewardship.

12.1 This treasure box teaches visitors how to make nature sculptures and encourages them to create their own. Photograph by Steven Glazer.

Avoiding Vandalism

Vandalism, unfortunately, can be a problem. In the Upper Valley region served by Valley Quest, 85 percent of the Quests are generally safe, sound, and out there waiting to be found. Approximately 10 percent of the boxes will be lost or vandalized on an occasional basis, say once every year or two. The final 5 percent, however, get heavily hit—some disappearing three, four, or five times in a single season. If the latter is the case, you most likely have made a bad decision regarding box placement. Sometimes the perfect place to put the box is also the place where teenagers hang out! Be sure to carefully analyze your hiding place site before publishing your Quest. You may find, as we have, that on occasion you will need to retire a Quest if you cannot guarantee the box's security.

Collecting Quests for Publication

If you aspire to collecting Quests from community members and publishing them, we recommend that you establish a standard submission form to insure that all of your Quests are complete. Local groups creating Quests can rely on this document as a checklist that all tasks are completed and in order.

Sample Quest Submission Form

Thank you so much for taking the time, energy, and creativity to create a Quest! We appreciate your generous offering to this community.

In order to make sure everything is complete—and to insure that your Quest is included in the next edition of our publication—we ask you to please take a moment to review the following checklist, answer a few questions, and return this form:

Checklist:

_____Original map art

_____Clues (that have been tested! On disk, please, or by email)

_____Compass rose, indicating north (if necessary)

_____Precise directions to the Quest's starting point:

_____Landowner permission (if required):

    Name _____ Phone _____

_____Estimation of time required to complete Quest (round trip): _____

_____Degree of difficulty: _____ Easy _____ Moderate _____ Difficult

_____Special features: _____ Architectural _____ Historical _____ Natural

    _____Vista (please check all that apply)

_____Walking conditions: _____ Indoor _____ Pavement _____ Trail

    Other _____

_____Optional gear: _____ Canoe _____ Compass _____ Bike _____ Binoculars

    Other _____

_____Season: _____ April–November _____Year-round (check one)

_____Treasure box complete: _____ Box _____ Stamp _____ Stamp Pad

    _____ Sign-in book _____ Other materials included: _____

_____Treasure box placed out there: _____ Yes _____ No

_____Quest box monitor: Name _____

    Address _____ Phone _____

    Email _____

"Cheater's directions"—i.e., the exact location of box for prompt monitoring or helping frustrated visitors find their treasure:

_____

PUBLICATION FORMATS

Our first Quests were on $8\frac{1}{2}$ by 11–inch paper, two to three sheets long, and stapled in the corner. The four main components of these early Quests were the locator map, the prose location directions, the clues, and the Quest map.

12.2 The Valley Quest map books from 1996, 1997, 1999, and 2001. The number of Quests grew from twelve to eighty-nine over the course of five years. Photograph by Simon Brooks.

With the second generation of Quests, we added a decorative border and played up the significance of the compass rose. The revised format made the name of the Quest and the name of the group making the Quest more prominent and added an estimation of how long it takes to complete the Quest.

In subsequent editions of the Quest book, we came up with a format of one Quest per spread (each Quest on two facing pages). While some Quests won't work this way because they are too long, this approach seems the most favorable. With all of the components of the Quest in clear view in front of the visitor, it is easier to remember to make use of both the map and the clue elements. Often, on multipage Quests, the map gets overlooked. Most likely, you won't have a whole book of Quests for some time, so your initial considerations should concern the creation of single publications that can eventually grow into a series.

Single-Page Quest. All of the key information is on one page, with one side offering Quest map, compass rose, and clues, and the other side answering such questions as "Who made this Quest?" "When?" "How long does it take?" "Is it easy or difficult?" "How do I get there?" "Do I need to bring anything else?"

Welcome to the Green Trail Quests!

Green Trail Quests is a series of treasure hunts that invite you on walks to explore the natural, cultural and historical attractions of South County. Located along the Rhode Island Green Trail, some sites are well known and others will be a new adventure. By following a map and deciphering clues, each Quest will lead you to a special place. At the end of each Quest is a hidden treasure box with information about the site, a unique rubber stamp and a log book. Record your visit in the log book and collect an impression of the stamp as a momento of your success in completing the Quest and finding the treasure box. Questing is a recreational activity developed by Antioch New England Institute which is becoming popular throughout New England.

Finding Your Way

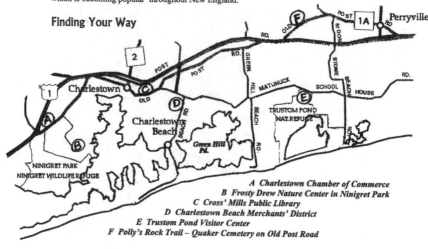

A *Charlestown Chamber of Commerce*
B *Frosty Drew Nature Center in Ninigret Park*
C *Cross' Mills Public Library*
D *Charlestown Beach Merchants' District*
E *Trustom Pond Visitor Center*
F *Polly's Rock Trail – Quaker Cemetery on Old Post Road*

Green Trail Quest Code

Keep to established paths. On roads, stay well to the side, away from traffic.
Avoid damaging fences and stone walls. Respect wildlife, plants and trees, and don't litter.
There's no need to dig—none of our Quest boxes are buried underground.
Please respect the treasure box as the private property of the host group.
Don't reveal the location of hidden boxes and please re-hide the box carefully where you found it.
Obey any posted regulations regarding dogs, and always keep dogs under control.

For Information and Other Quest Maps

Contact Green Trails Quest Central at the Charlestown Chamber of Commerce on Old Post Road in Charlestown, RI. Phone: 401-364-3878.

Brought To You By

Green Trail Quests are developed by the Salt Ponds Coalition, the South Kingston Land Trust, the Charlestown Public Library, the Charlestown Chamber of Commerce, the Friends of the National Wildlife Refuges of Rhode Island, the Rhode Island National Wildlife Refuge Complex and the National Park Service Rivers & Trails Program.

12.3 Green Trail Quest. Most programs begin with one-page Quests. All of the necessary information appears on a single $8\frac{1}{2} \times 11$ sheet of paper, with map and clues on one side and driving instructions, Quest code, and other relevant information on the back, as shown.

You may want to include fun local facts as well. If you three-hole-punch single-sided Quests, people can begin to collect them and store them in a three-ring binder. This is a very cheap and effective publication method.

A related approach that has worked well for some groups is to print a large number of template sheets in colored ink. The back of the sheet is the same for all of the Quests in the program and includes information about the Quest code, the sponsors, the history of Questing, Quest contacts and a locator map for all of the Quests. The front is blank, except for a simple attractive border and the Quest program name. Individual Quest maps and clues can be inexpensively copied in black ink into the blank space. These sheets share a common graphic identity and can be displayed together.

Front/Back Half-Fold. A single sheet of $8\frac{1}{2} \times 11$ or legal-sized paper can be folded in half to make a pocket-sized four-page booklet. This format offers extra space that can be used for a decorative cover, a how-to-get-there map, information about your program or the group that made the Quest, or additional facts about your Quest site.

Ledger Booklet. Ledger paper is wonderful in that it enables you to increase the size of all of the elements: you can use a bigger font for your clues, making them easier to read, and you can make your map larger and clearer for both younger and older eyes.

Two-Page Ledger Booklet. Developed by the Marion Cross School in Norwich, Vermont, this has quickly become the preferred product for school groups creating Quests. Why? The booklet allows the text and map to be larger, as mentioned above. Groups can add a decorative cover on the front and describe their group and the project on the back. There is also a lot of "white space" that can be filled with art, additional information, footnotes, and credits.

If you have money available for production, or if you have graphic arts expertise available, the scope and execution of your publication can increase dramatically.

DISTRIBUTION

Once you have decided on your publication style, you'll need to figure out how you are going to get the word out so people can use the Quest. If you are making a Quest with a small group, at minimum you are going to want to send

Hidden Treasures? Lost Cities? Valley Quest!

By Steve Glazer

Hidden in the southeast corner of Vershire – just along the West Fairlee town line – is a lost city. Formerly the site of the Ely Mine, this community was once bursting with a population of more than 1,000. The village had two churches, a school, a store and a post office, with extensive ox-cart traffic hoofing carts back and forth to the train depot in Ely. Today, however, if you drive by quickly and don't look closely, all you will see are trees.

Lurking just out of sight, however, along both sides of Beanville Road (also known as South Vershire Road) are cellar holes, stone walls and old, tell-tale apple trees...along with many, many signs of a time now passed. Follow the accompanying *Valley Quest* map and clues of the

Copperfield Town Quest to discover a treasure and get the whole story for yourself.

This particular adventure, the Copperfield Town Quest, was created by Barbara Griffin's 2nd and 3rd grade classes at Vershire Elementary school last spring and is one of 89 Quests in the new book *Valley Quest: 89 Treasure Hunts in the Upper Valley* (Vital Communities, 2001, $12.95). The Quests stretch across 51 Upper Valley towns. *Valley Quest: 89 Treasure Hunts in the Upper Valley* is available at fine book stores and retailers across the region, or call (802) 291-9100 to reserve your copy.

Steve Glazer is the Valley Quest Coordinator at Vital Communities of the Upper Valley.

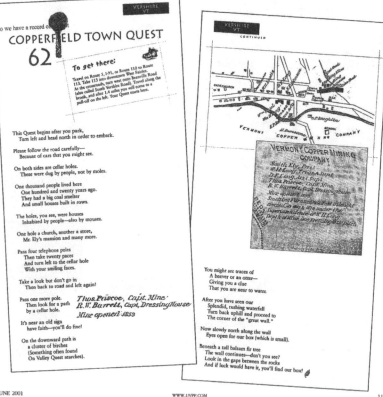

12.4 Local newspapers are often pleased to publish your Quests as regular feature columns or articles. The *Upper Valley Parent's Paper* publishes a Quest in each issue during the spring, summer, and fall seasons.

each group member home with a copy of the finished Quest. Better still, you could send them home with several copies, so these can be passed on to neighbors and friends. If your school or community has a regular newsletter, the Quest can be submitted and published in it. Our community, the Upper Valley, has a monthly "parent's paper." We have had good success releasing the Quests serially through the spring, summer, and fall issues. This method puts the Quest right into the hands of our target service population. In our neighboring Monadnock region, the local daily has included full-page inserts of Quests and accompanying activities.

Quests that make use of a trail can be placed in a box at the trailhead; people visiting the trail for the first time can find them there. Quests can be placed in welcome or visitors' center kiosks. We developed a Quest for the White River Junction Visitors' Center. As the center was located adjacent to the train depot and its décor reflected the theme of transportation, we developed a display using a stack of old steamer trunks. The top trunk was fitted with a rack offering complimentary copies of the Quest. The Green Trail Quests program in Charlestown, Rhode Island, distributed their Quest series through a "Quest Central" rack in the chamber of commerce visitors' center. Additional copies of all of the Quests are available at each starting point.

The World Wide Web

Quests can be distributed by making use of the latest in global information technology. Develop your own website, or use the website of your city, neighborhood school, or regional organization. Vital Communities offers a "Quest of the Month" on its website, <www.vitalcommunities.org>, and Antioch New England Institute regularly publishes recent Quests at <www.schoolsgogreen.org>. Letterboxing USA offers links to treasure hunts across the states. You may choose to send your Quest to the Letterboxing USA webmaster at <www.letterboxing.org>.

Publicizing Your Quests

Many groups celebrate the completion of their Quest by holding a Quest Fest. We like to enhance these with the music of a local band, arts and crafts tables to make stamps and passport books, childrens' games, and ice cream sundaes. Related local nonprofit organizations find these events a great place to display their materials as well. We've found that these kinds of intergenerational community events are appealing to the local media. Send a press release two weeks in advance and follow up with a phone call or email.

12.5 Beyond publication, we recommend that you launch your Quest with a community celebration. Photograph by Jon Gilbert Fox.

A Quest Season

Each Quest season in the Upper Valley begins with a series of potlucks two weeks before the season traditionally opens on April 22, Earth Day. The potlucks are opportunities for box monitors and volunteers to gather and celebrate a new season. All of the boxes are refurbished: missing stamps replaced, new sign-in books added, ink pads cleaned up, and so forth. Parties can also be held at the season's close, with a harvest festival around Thanksgiving.

Use and Impact

The more limited your distribution, the fewer users you will have, which may or may not be a good thing, depending on your intentions. Be clear about your goals, and please remember that some secret places should remain secret. Now that the Valley Quest program has a published book of Quests, *Valley Quest: 89 Treasure Hunts in the Upper Valley,* and that book is available in stores, local school libraries, and regional public libraries, most treasure boxes get between 40 and 250 hits in a season. The lower number reflects the more remote locations, while the most popular Quest received 500 hits during the 2002 Quest season. (This program serves a sixty-town, 800-square-mile region with a total population of about 150,000 people.) Within this region we might have an audience of perhaps 4,500 Questers, or 1,000–2,000 questing families. The average number of visits per box has climbed dramatically since our first publication, growing from forty in 2000 to seventy in 2001 and one hundred in 2002.

ADMINISTERING A QUEST PROGRAM

For a Quest Program to be successful, there need to be a few really good Quests; the boxes need to be maintained; the Quest maps need to be distributed; and people need to use and enjoy them. Ultimately this is a community-based activity that can be undertaken with just a small group and very little investment of cash. The scale of your Quest program will depend on your goals. Is your emphasis on Questing as a recreational activity? Questmaking as a community-building activity? Questmaking as an educational activity? Or all of the above?

Recognition

We've found that honoring the accomplishments of Questers who sustain their interest and stamina long enough to visit numerous boxes goes a long way toward building an engaged community of Questers. In Dartmoor, if one submits proof of having visited one hundred boxes, and a small sum, one can receive a fabric patch. We set the bar a little lower, at twenty boxes, and have found that patches and certificates have become two key elements of our Quest program. Those who take the time to earn a patch will see the true value of your program:

8/30/00

Dear Valley Quest Folks,

My daughter, Libby (age 6), and I had a great time this summer hiking Quests. We have lived here for three years and never knew there was a pretty and public lake in Orford or that anyone could hike into the woods in so many places and see such wonderous stuff. Cool beans! Enclosed is a copy of our stamp book.

Thanks for the guide,

Chris

12.6 Quest patch and certificate. Photograph by Simon Brooks. Patches and certificates are awarded to people who successfully complete twenty Quests. The Valley Quest patch was designed by Susan Clark.

Your goals are met, too, as people deepen their connections to your community. Parents and children earn patches; grandparents and grandchildren earn patches; elderhostelers earn patches; visitors to your community earn patches too.

Other programs have used less expensive pins and stickers to accomplish the same goal. We also award certificates of appreciation to individuals and groups who take the time to create new Quests. The goal of all of this sharing can be summarized in the sign-in below:

10/01/01

What a beautiful day here on top.
What colors!
It is so quiet here you can
Actually hear the wind.
Used to come up here a lot when young.
There never seems to be enough time.
Please keep this place just like it is.

13) Building a Quest Program

WHEN DAVE MONK, director of the Salt Ponds Coalition in Charlestown, Rhode Island, decided to start a Questing program, his first step was to invite a number of local organizations to consider joining him in linking Questing to the region's Green Trail project. Dave's goal in launching a Quest program was to draw some of the heavy tourist activity away from the delicate Salt Ponds Watershed near the Atlantic beaches. He reasoned that he would have a greater chance of success if he enlisted the participation of other allied organizations. Though some of the organizations he thought of had not worked together as partners before, he was able to convene an initial planning meeting with support from the National Park Service Rivers and Trails Program that resulted in six organizations signing on as Quest program partners.

Each organization had its own reasons for joining the Questing program, not one of which including protecting salt ponds. Nevertheless, their goals proved to be compatible. The South Kingstown Land Trust saw Questing as a way of attracting visitors to their protected properties and building public support for land conservation. The U.S. Fish and Wildlife Service Rhode Island Refuge Complex saw Questing as a means of providing visitors to Trustom Pond National Wildlife Refuge with quality educational programming during a time of budget restraints on hiring staff interpreters. The Charlestown Chamber of Commerce saw Questing as a way to attract beachgoers to the shops of local merchants. And the Cross Mills Public Library saw Questing as a way of deepening public understanding of local history.

Questing proved to be a successful means of combining efforts and working toward each of these separate goals. The partnership resulted in the creation of six Quests, which were then made available to the public at "Quest Central" in the chamber of commerce tourist information center. Green Trail Quests received an Excellence Award from the South County Tourism Council in their first summer.

Your interest in organizing a Quest program in your community might arise from a desire similar to Dave's—or one of his partners. On the other hand, you might simply find that the creation of one Quest by a teacher or resident sets a trend in motion that seems promising. At some point, it might make sense to pull together a group to think about how Questing can strengthen your community and help it to meet its goals.

At these early organizational meetings, discuss your motivation for beginning a Questing program. Examples from other Questing programs around the country include

- generating more visits at a municipal park system;
- having students engage in community research and service;
- increasing ecotourism;
- promoting the conservation of natural areas;
- boosting public support for area farms;
- preserving a city's character;
- celebrating the ethnic traditions of a neighborhood; or
- making a community an even more fun place for a visit.

Dick Norton, a retired doctor, first went on a Quest six years ago while visiting his son in the Upper Valley. He was enchanted with the experience and contacted Vital Communities to learn more. Here he describes the creation of the South Shore Quests program:

It took about six months from the time I first went on a Quest in Lyme, NH, to the time I called an evening meeting at our house in Hingham. There were fourteen or fifteen friends and acquaintances that I thought might be interested in starting a Quest group. I had told each of them about the Quest movement, and I had on hand maps of the South Shore and the Town of Hingham. I also had a Tupperware box, a stamp pad, and assorted stamps.

I spoke briefly about the history of Letterboxing. They listened respectfully and asked thoughtful questions. I asked if they were ready to put together a committee, and enough people were in favor that I called for volunteers for each of the officers. From that time until now we have functioned fairly smoothly. There was some attrition in the first few months, and we filled the vacancies easily. Our South Shore Quests Committee likes to meet monthly, although we sometimes skip a month in the mid-summer.

The Boston Chapter of the Appalachian Mountain Club gave the South Shore Quests group our initial funding so we could purchase supplies and get our first booklet of four Quests printed. They gave us a little more the second year, by which time we had nine Quests, and a few of the committee members added to the treasury. Over the last few years we have been able to maintain a reasonable back balance without further fundraising. In other words, we print and distribute the booklets of twenty Quests with enough margin to cover the costs. We carry on all our activities on a volunteer basis, which leads to some inefficiencies, of course.

About a year ago there was the usual addition of new Quests to our list, and the question arose as to whether to let the total rise from twenty to thirty or more. The committee felt very strongly that we could manage twenty but not more so we have maintained this as a top figure, and we have retired some Quests each year to make way for new ones.[1]

The South Shore Quests program now includes Quests hosted by a state park, a municipal park, and the town landing, among others.

PARTNERSHIPS

Once you get started, you'll find that there are many partners who might cosponsor a Quest site, or even an entire Questing program: museums, arboretums, land trusts, chambers of commerce, historical societies, environmental organizations, arts and cultural organizations, conservation commissions, and youth groups such as Scouts, 4-H, and Boys and Girls Clubs.

There are special interest clubs and organizations that might like to design Quests that share their passions—a quilters' Quest, for instance, or a snowmobile Quest, playground Quest, birdwatchers' Quest, herb garden Quest, even a winetasting Quest. If you're lucky enough to have a national, state, or provincial park in your community, it can make a fine partner. The same goes for municipal parks and protected areas and private nature centers and preserves. Businesses might choose to become involved in your Quest program, too. Dick Norton's South Shore Quest newsletter includes advertisements for a local rubber stamp shop. Schools, as previously discussed, are natural centers for Questing activity.

If partnering is relatively new to you, you can get a strong start by gathering your group around a flipchart or blackboard with three columns marked on it. Work together to brainstorm ideas in three categories. In the first column, list all of the *potential partners* you can collectively think of. Look beyond those you already know well. For each of these organizations, list in the second column what you guess to be their *primary focus*. In the third column list the way that Questing might help each of them *address* or achieve their goals.

Before you approach other groups to cosponsor a Quest, test your assumptions about their primary interests by searching some of their publications, including their mission statement. If you sense that there may be some compatibility, arrange for an exploratory meeting, perhaps a one-on-one chat

December 14, 2001

Steve Golden
National Park Service
15 State Street
Boston, MA 02109

Dear Mr. Golden:

This is to confirm that the Beebe School is interested in receiving the services of Antioch New England Institute to help develop a Quest for Fellsmere Pond/Park. This local pond is part of the Mystic River Watershed and the center of many environmental and community studies at our school. We consider Fellsmere Pond our "Outdoor Classroom" and work to encourage stewardship of this valuable community resource. Service-learning projects are strongly endorsed here and the development of a Quest for our community fits into Beebe's educational philosophy.

We have the commitment of three fifth grade teachers to incorporate developing a Quest into their curriculum. We have an ongoing relationship with Antioch New England Institute and are pleased to extend our relationship to include this Quest development.

If you have any questions or comments please do not hesitate to contact me. Thank you for giving Beebe School this wonderful educational opportunity.

Very truly yours:

Robin W. Jorgensen

Cc: Peter Magner
     Charles Tracey
     Delia Clark

13.1 Support from the National Park Service for a partnership between the Mystic River Watershed Association and a local elementary school proved to be the perfect way to establish some Quests in an urban park near Boston.

over a cup of coffee. If there's interest, arrange for a larger group meeting to consider moving forward.

REACHING OUT

Why not take advantage of Questing's universal appeal to build bridges between the various constituents in your community—and reach out to visitors as well? One way to do this is by translating your Quests into the various languages spoken in your community. When the Hulbert Outdoor Center invited a group of New York City families who had been affected by the attacks on the World Trade Center to spend a week relaxing in their rural lakeside camp, they faced the challenges of a language barrier. One solution that made their Spanish-speaking guests feel welcome was translating the Fairlee Glen Falls Quest into the Cascadas de Fairlee Glen Quest:

> Cuando encuentres el segundo
> Marcador anaranjado en tu paseo,
> "De una vuelta a la derecha y camine
> 30 pasos," eso es lo que decimos.
>
> Ahorra busque una raiz ascendente,
> Que vista mas Hermosa
> (Como la mano di un gigante
> Pero por favor no tengan miedo).
>
> Mirando hacia las altas cascadas,
> Una cepa inclinada aparecera a su derecha.
> Camine 22 pasos y mire hacia abajo
> Para descurir nuestro sitio donde
> Queda Caja de Tesoro de Quest.

When the Quebec Labrador Foundation/Atlantic Center brought a group of rural women leaders from the Czech Republic, Slovakia, and Poland to the Marsh Billings Rockefeller National Historic Park, they were looking for more than a tour. They wanted these strong women—on whom so much of their community's future depends—to have some first-hand experiences with community-building strategies used in the United States. Thus, the Village Green E Quest was translated into Slovak:

> Zostaňi na ceste Route 4, ktorá so stráča do dialky
> Aš kým neprídeš k budove, ktorá je úplne biela.
> Neboj sa, neplač, budeš vedieť, že si tam,
> Keť uvidíš túto bielu budovu ěnieť do výšky.
> Pojet rounakých dverí na prednej strane budovy = ___(B)

MAINTAINING YOUR PROGRAM

The long-term success—and impact—of your Quest program depends on its ability to attract participation. Participation in turn depends on the promotion, whether formal or word-of-mouth, and maintenance of your Quests.

We've found that promotion is best accomplished through a cross-media approach. Seek local newspaper, TV, and radio coverage by creating time-sensitive events such as launching parties at the beginning of each new season. Develop human interest stories with each new Quest submission that focus on new Quest sites, Quest content, or your partner groups. Develop a newsletter or website that shares your newest Quests, updates old ones, and includes information of interest to your community. "Tally Whoa!" in the South Shore Quests newsletter, shares the number of visits to each box in a given season. You can also share stories from the field, photographs ("Where am I?"), the names of new patch earners, fundraising updates, and quotes from your treasure box sign-in books.

Nurturing a Culture

Treat your volunteer box monitors well. Publicize their names, keep in contact with them, and thank them often. Hold parties in their honor where they can gather and swap stories and tips. Make it easy for them to communicate with you about any problems. You might even choose to equip them with a treasure box "first aid kit": a replacement stamp pad, logbook, temporary stamp, and box label.

You might, as we did, consider forming a new nonprofit organization for your Questing program, though a simpler route would be to find an existing nonprofit sponsor or partner with a compatible mission. As it grows, your program will develop more direct expenses, such as the printing of your Quest booklet and newsletter, box supplies, advertising, and the like. You may grow to the size that allows you to hire a part-time program coordinator.

Fundraising

One of the best ways we've found to raise money to support the basic needs of a Questing program is simply to Quest, and share the fun of mystery, maps, clues, stories, and hidden treasure with those from whom you're requesting money. Invite potential contributors out for a ramble with you, or create a Quest as part of a fundraising event. Follow Dartmoor's example and host a Ten Box Trot fundraiser, featuring checkpoints at some of your actual Quests and at others created just for the day. Local community foundations, and regional arts, education, and environmental foundations have all shown strong support for Questing.

Jayni and Chevy Chase have been supporters of environmental education initiatives across the country, contributing not only funds but also entertain-

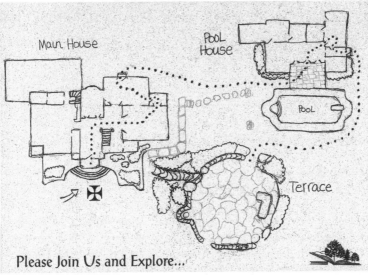

13.2 a and b Invitation to a Quest-based fundraising party. Map by Cydney Chase. Take advantage of the appeal of Questing to build support of your program.

ment industry know-how. They applied their well-developed sense of theater and fun toward a fundraising party they hosted at their home for the Center for Environmental Education (CEE), which coordinates Antioch New England's Questing program.

The invitation featured a map of their home and grounds drawn by their daughter Cydney. There was a mysterious dotted line wandering here and their and a message inviting people to come and explore. As guests arrived at the event, a team of youthful guides greeted them, handing each visitor a crisp new passport book. Guests followed the map and clues from station to station, around the house, yard, and pool. Each stop on the Quest featured a presentation about a different CEE program or initiative, and each presentation closed with the adding of a stamp to the guest's passport book. The Quest ended on the terrace, where evidence of a full passport and collection of stamps bought you a drink, some refreshments, and the chance to enjoy Chevy playing the keyboards.

Dave Monk also found a creative way to convey the fun of Questing in his request for assistance from the National Park Service Rivers and Trails Program. Dave wrote his grant application entirely in rhymed verse, addressing the application form's technical questions with a series of limericks. Here are some excerpts:

How will the project support your conservation efforts?

Some Rhode Island groups near the ocean
For beaches and ponds have devotion,
But tourists galore
Come visit our shore
Which causes environmental commotion.

On Green Trail are plenty of places
Where Questers can go through their paces.
It's a place to escape
The crowds at the Cape
And ramble through wide open spaces.

How will you promote public use of your project?

The South County Vacation Planner
Tells people where they can get tanner.
Here we'll show them the way
They can Quest for a day.
We'll reach oodles of folks in this manner.

The Chamber of Commerce Guest Center
Has twenty five thousand that enter
Each year to find out
What fun stuff we tout
So here we'll put Quests front and center.

Who are the key cooperators in your project and what is
their level of commitment?

There currently are seven teams
Whose members have similar dreams
To figure out searches
Near woodlands and churches
And libraries, stone walls and streams

They'll volunteer much time and zest.
Their clues will give Questers a test.
They'll draw all the lines
And put up the signs
Saying Welcome to Green Places Quest!

What are the specific conservation benefits of your project?

Our region has country bucolic
Where many a Quester can frolic,
But we want to combine it
With a green business climate
To help manage growth economic.

Green Places support a tradition
Of more open space acquisition.
As Questers cavort
They'll surely support
The benefits gained from this mission.

How will you raise funds and project support?

Each of our Quest groups commit
To volunteer people most fit
To make one or more
Great Quests to explore.
This will lower the costs quite a bit.

As the Chamber of Commerce expands
Their Visitors' Center is planned.
There'll be Quest information
In a central location
With Questing advisors on hand.

We've volunteer work to invest
In a project we don't take in jest,
But it sure would be great
If you'd help us create
"Welcome to Green Places Quest"![2]

We are pleased to report that Dave and his partners got the support they were looking for.

14) From Questing to Stewardship

IN OUR WORK WITH TOWNS AND CITIES across New England, we've found that healthy, strong, vibrant communities emerge directly from the energy, passion, and good humor of the people who live in them. We've found, likewise, that the motivation to participate in community affairs—and act as stewards of local natural and cultural heritage—comes directly from a strong sense of place: a bone-deep understanding of and concern for the landscapes and people of our home ground.

But where does sense of place come from? In our experience, love of place—as with love for another person—is an emotion charged with magic and mystery. Sometimes it arises fast and furious, other times in a slow upwelling. But always love arises from intimacy: cherished details, shared experiences, the admiration of specific qualities, and a true concern for wellbeing. Once crafted, a sense of place can endure all but the harshest blows.

Through coaxing people out of their home cocoons and into the heart of their communities and the natural world, Questing supplies some of the building blocks that can lead to love of place. Through Questing, we might learn where the delicate wood anemones first bloom in spring, or that the first globe ever created was made right here in our own town, or that the older gentleman down the street—the one we've always written off as a loner—is actually a funny and wonderfully talented mapmaker. As we visit places again and again, we notice their subtle nuances and find ourselves, quite naturally, wanting to share them with others.

Wendell Berry has written that community revival "would have to be a revival accomplished mainly by the community itself. It would have to be done not from the outside by the instruction of visiting experts, but from the inside by the ancient rule of neighborliness, by the love of precious things, and by the wish to be at home."[1] Questing, as a tried and true arrow in the quiver of place-based education techniques, is a powerful approach to community revival and the development of ecological literacy. By simultaneously addressing the three principal goals of place-based education—community vitality, environmental quality, and lifelong learning—Questing offers a practical pathway toward community stewardship.

The direct outcomes of launching a Questing program, whether large or

14.1 Exploring the treasures of your community is a step toward stewardship. Photograph by Jon Gilbert Fox.

small, are many. We've learned that the development of sense of place—and stewardship behavior, too—is a transferable skill, useful again and again in our mobile society. The research of child development experts has shown that middle childhood, as well, is a sensitive period in which free play in semiwild areas and role-modeling by adults play crucial roles. In this time of heightened awareness and sensitivity, environmental exploration can have a tremendous influence on the development of lifelong civic and environmental values as well as have a sense of self-efficacy. The opportunity to come together in a common purpose, and to play outdoors in the landscape and neighborhoods with adults who care, is both healthy recreation and potent learning experience.

Social capital, that invisible web of reciprocity and trusting relationships that binds communities together, is best developed through informal opportunities where community members come together for work and play. Participation with other community members in leisure activities and collective civic endeavors acts as an inoculation against damage from the swirling, overwhelming forces of globalization. Insofar as it strengthens social capital and builds long-lasting community partnerships, Questing serves as a kind of prevention work for healthy communities.

Questing is also a way to develop an informed electorate that appreciates the value of open space and historic preservation. If community members spend time exploring public lands and special places in their own and neighboring communities, they will be more likely to vote for conservation land acquisition or historic preservation. They will be more likely to see the preservation of their home ground as a personal and loving responsibility.

Out Questing, we work and play together, shoulder to shoulder, with a diversity of neighbors, launching new friendships as well as cementing old ones. We learn that Dorothy was right, "There is no place like home," as we find and tap the inner resources to leave a legacy of healthy environment and rich culture for those who come to these places after we do.

APPENDIX: THE VILLAGE QUEST CURRICULUM

The Village Quest project is a six-to-eight-week integrated curriculum unit that uses a community village to teach local history. It can be modified to serve students in grades 3 through 8. We offer it as a model to help support the use of Questing in school curricula.

INTRODUCTION

Throughout the rural areas of this country are vestiges of the small hamlets that once sprinkled the landscape like so many freckles. Each township, within its (often) rectilinear perimeters, contained from three to two dozen small settlements, recognizable now by perhaps a small cemetery surrounded by tall trees, some buildings at a crossroads, an old schoolhouse or church, or perhaps simply a collection of cellar holes in the woods. Why all of these tiny villages? How did they come to be settled? How did people live and work in them? How did their cultures and habits affect their habitat—their community and place, the fields and the forest? Why and how is it that these villages disappeared or were transformed?

Through investigating remaining buildings, reading old maps to find "ghost" sites, examining town records and tombstones, and listening to the stories related by older residents about their lives, students can learn much about the history and life of their community.

In order to help students cultivate an appreciation for and understanding of the small communities they may pass through or live in, we created the Village Quest. It is best suited for a self-contained classroom but has also been implemented as an elective. The lessons/activities are linked to specific standards, mainly in history and social studies, with the Vermont Sense of Place Civic/Social Responsibility Standards as an umbrella over the entire project. As written, the eight lesson plans are designed so each may be completed in one or two ninety-minute periods.

Each step of the unit contains objectives, procedures, a materials list, informal assessments, and either follow-up or extension activities. Follow-ups are essential to the completion of the project, while extensions are suggestions for enrichment. The final rubric model offered can be used or adapted to assess student performance on the entire unit. Since each Quest, no matter how similar in theme, is different—based upon the available historical sources, the interests of the students, and the character of the village itself—the lesson plans are meant to be guiding principals rather than prescriptions. However, the lessons do provide the skeletal structure required to produce a Quest that fulfills the overall intent: to foster in the students an awareness of and appreciation for the environmental and cultural heritage of their community.

This unit will result in a Quest that falls into the Quest taxonomy described in chapter 5 as:

➤ A-1 or A-2\—Basic Clue to Clue or Elusive Clue to Clue
➤ B-1\—Full Map and Clues Set
➤ C-5—Cultural History Quest

You can modify your work with the unit to bring in any of the other models, however, depending on your interests and goals. For example, if you want to include some math standards in the unit, you might choose to modify the unit to create a Number Hunt Quest focused on the dates of the buildings.

Special thanks to Marguerite Ames, Becky French, Amos Kornfeld, and Ros Seidel for assistance with developing, writing, testing, and refining this unit.

OVERVIEW OF LESSONS

Lesson I. Introducing the Idea of Village Settlements
Lesson II. Investigating the Village
Lesson III. Drawing the Village and Landscape
Lesson IV. Primary Source Investigations at Municipal Office, Local Historical Society, or Library
Lesson V. Oral History Interviews
Lesson VI. Writing the Quest: Clues that Teach and Move
Lesson VII. Pulling it All Together
Lesson VIII. Testing Your Quest

PREPARATION

Before you begin working with your students, do some preliminary research.

Essential Question: What is the story of this community? What causes, conditions, actions and attitudes led to the rise and fall of a "Vermont village?"

Standards & Evidence	Criteria	Learning Activities	Products & Performance	Assessment
OVERARCHING STANDARD	Demonstrate knowledge and history of the local environment and how the community relies on its environment to meet its needs.	1. Introduce Valley Quest		
Understanding Place **Students demonstrate understanding of the relationship between their local environment and community heritage and how each shapes their lives.**		2. Introduce village settlements	Map of community settlements	Answer Key – Teacher's Map
	Describe the role geography and industry played in the development of their local community over time.	3. Investigate a village	Map of village settlement and overlay Journal entries	Answer Key – Teacher's Detail Checklist
		4. Draw the village and landscape	Drawings of village / detail drawings	Task Specific Rubric – "Drawing"
	Demonstrate knowledge of the changing face of their community and culture: settlement, the rise of industry, rural abandonment and facets of our contemporary, regional culture.	5. Primary source investigation conducted at historical society, town office, or local library	A) Complete answers to "target questions" at historical society	Answer Key
			B) Create a question that can be answered using a primary source found at the historical society.	Task Specific Rubric – "Research"
OTHER STANDARDS ADDRESSED:			C) Share an anecdote "discovered" at the historical society.	Task Specific Rubric – "Research"
Continuity and change. Demonstrate an understanding that perceptions of change are based on personal experiences, historical and social conditions, and implications...	Describe their community in terms of continuity and change? What elements of their community have stayed the same? What are some of the more dramatic changes that have taken place?	6. Oral history gathering	A) Develop question(s) and follow-up question(s) for oral history gathering.	Task Specific Rubric – "Research"
Movement & settlement. Analyze and evaluate the causes and effects, processes and patterns of human movements, both chose and forced			B) Create written summary of information gathered through oral history	Task Specific Rubric – "Research"
Being an historian Collect and use primary sources Use oral history methods	Apply knowledge of local environment through active participation in local projects	7. Writing the quest	A) Complete "movement" clue	Task Specific Rubric – "Clues"
			B) Complete "teaching" clue	Task Specific Rubric – "Clues"
		8. Culminating activity: producing the Valley Quest treasure hunt	Complete 1 or more Quest elements: compass rose, map, stamp, box, etc.	Task Specific Rubric – "Teamwork"
	Collect and utilize primary sources.	9. Testing the quest with community members		
	Collect and utilize oral history.	10. Taking it public: distributing the Quest to the community		
		Lessons lead up to the production of a Valley Quest Treasure Hunt that teaches the role geography, environment and industry played in the development of the community; demonstrates how the community has stayed the same and changed over time; and the relationship of those changes to changes in town, regional and/or national culture.	**CULMINATING ACTIVITY** Students will demonstrate their learning through the "product" of the quest, and the successful completion of all products and performances.	Student's mastery will be assessed via the Evaluation Rubric and a review of the portfolio of completed products. Products will be assessed via the above.

A.1 This five-column chart provides an overview of the entire unit.

Local History

Try to locate a published history for your community or neighborhood and begin to sift through it. For example, one small town in Vermont has three:

Mills and Villages of Thetford, Vermont by Charles Hughes
A Short History of Thetford, Vermont by Charles Latham Jr.
Tales of Thetford by Helen Savery Paige

Historical Maps

See if you can find at least one historical map of your town, city, or village, for example, the *F. W. Beers Atlas of Windsor County*, 1869.

Historical Society

Locate a local historical society, museum, or study group. If they have a collection of materials, ask for a general tour focusing on the different primary and secondary sources housed there. This will help you to discover more quickly some of the key stories—natural and cultural—of the community, while providing reconnaissance regarding materials that can offer your students the opportunity for first-hand discovery and learning experiences.

GETTING STARTED WITH YOUR STUDENTS

Ideally, before you initiate a new Quest you should take students on an established Quest to give them a sense of what one can be and to help them understand the two fundamental types of clues, those that teach and those that move visitors from one site to another. As you move through the Quest, have students consider which clues are which. At the end of the Quest, encourage them to assess the quality of the Quest. Was it fun? Was it successful in getting you from place to place? Was the information interesting and the place worth visiting? Did the map help? Review the basic elements that make up a Quest: the map, the clues, the treasure box, the stamp, and the sign-in guest book.

If it is not convenient to go on an existing Quest, then go on a simple, in-school treasure hunt (see Chapter 6) to give students a sense of what a Quest is. You can easily make a basic sample Quest in less than an hour, or you can have them read through a sample Quest from this book or visit the "Virtual Quest" at <www.vitalcommunities.org> or <www.schoolsgogreen.org>. It is important for them to understand the key components of a Quest: the clues, the map, and the treasure box. If the students have a clear sense of the goal when they start the unit, they will take on the activities with a greater sense of purpose.

LESSON I. INTRODUCING THE IDEA OF VILLAGE SETTLEMENTS

Focusing Questions

➤ What makes a village?
➤ Why do villages grow in certain places?

Standard(s)

Traditional and Social Histories: Demonstrate an understanding of the relationships among community leaders, important events, and people's lives.

Movement and Settlement: Analyze and evaluate the causes and effects, and processes and patterns, of human movements, both chosen and forced, in the community, state, and world (e.g., the impact of transportation or technology).

Understanding Place: Explore the interrelationship between the local environment and the local community heritage (e.g., settlement patterns, tourism, hunting, agriculture).

Length of Time Needed to Complete

90 minutes

Procedures

1. Have students work together to create a basic map of their town or community. You can do this either on the chalkboard or on individual pieces of paper. What is the community's shape? What towns lie to the north, south, east, or west of it? What are the community's main watercourses— and thoroughfares? Are there any key geological features? How many villages or settlements are there? (Note: this map can serve as a good pre-assessment.)

2. Lead a discussion framed by the following guiding questions:

   ➤ If you were the first settler in your community, where would you settle? Why?
   ➤ When settling previously unsettled areas, why might a village "grow" in a certain place?
   ➤ What might settlers look for in terms of geography and location when choosing a home site?
   ➤ What architectural or humanmade elements and natural resources, at a minimum, would you say it takes to "make a village"? Generate a list of core components.

3. Research Activity: Distribute ledger-sized photocopies of a historic map of your community. Have students locate watercourses, geological features, old roads, places where buildings are grouped, and community structures like churches, cemeteries, and schoolhouses. They can color code the natural features of the land: water in blue, roads in brown, mountains or fields in green. They can also color code manmade structures: schools in yellow, churches in pink, cemeteries in red, etc., and circle what appear to be settlements, i.e., clusters of buildings.

4. Drawing Conclusions: Do these clusters actually relate to natural features? How so? What elements are common to most clusters? Why are there so many settlements and schools in just one town? Why is this no longer the case?

Materials Needed

Hand lenses or loupes
Colored pencils
Oversized (ledger) photocopies of early maps (one per student)
A few contemporary maps to use for comparison (one per table)

Assessments

Product: Students have produced clearly colored maps, showing they recognize settlements, school districts, and key geologic and cultural features. Tool/reference point: teacher's map/answer key

Performance: Their concluding discussion reflects an accurate/reasonable interpretation of the data. Tool/reference point: checklist

Extensions

Comparing old and contemporary maps, students attempt to find the location of their home site on the older map as well as determine the local schoolhouse they might have attended. (This can be done in class or as homework.)

Extra credit: Students walk to that school or the location where it was and make a map of their walk, attempting to record the time and distance from their home to the school. Ask them to determine the time and distance from their home to their current school as well, for comparison.

LESSON II. INVESTIGATING THE VILLAGE

If there are multiple settlements in your community, and you have access to a bus or van, you may choose to take a tour of town and try to locate evidence of some or all of the old settlements. At each site you can make drawings or take photographs, make notes, or plot GPS points. The signs of settlement might mean a cemetery, an intersection, or a cluster of buildings. After that outing, you (or your students) can choose one settlement to focus on for the remainder of the project. Save the other settlements for future years. Alternatively, you can pre-identify a single village to visit and move directly on to the next activity.

Focusing Questions

> What are the key elements that make up a settlement?
> How can we find clues that tell us about the past?
> How has this (and our) community changed over time?

Standards

Geographical Knowledge, Sense of Place: Describe such patterns as population distribution, land use patterns, climate, transportation networks in Vermont, the U.S., and the world.

Movement and Settlement: Analyze and evaluate the causes and effects, and processes and patterns, of human movements, both chosen and forced, in the community, state, and world (e.g., the impact of transportation or technology).

Understanding Place: Demonstrate knowledge and history of local environment (e.g., soils, forests, watershed) and how the community relies on its environment to meet its needs (i.e., nutritional, economic, emotional).

Length of Time Needed to Complete

2 hours to a half day

Procedures

1. Opening activity, discussion, and guiding questions (in class):

   ➤ How many students attempted to find "their" schoolhouse? Who was successful?
   ➤ Was it still standing? If so, what was the current use?
   ➤ How far was it (time and distance) from students' houses?
   ➤ From our discussion last week, we determined that a village might have certain essential buildings. What were they? (Ask students to recall church, store, school, town hall, post office, cemetery.)
   ➤ Which of these buildings are still standing in the hamlet in question (where you might have gone to school)?
   ➤ What do you think the village was like the year your map was made?
   ➤ What is different about life in this community today?

2. Research Activity (in the field): If possible, locate a local historian (an amateur is fine!) to accompany you for this activity and provide interpretation. Walk through the main corridor of your chosen village, with each student holding an enlarged detail map that shows only the settlement in question. Bigger is better—legal or ledger sizes are the best. As they walk through the settlement, students make note of the elements from the old maps that are still there; they note what is gone (the "ghosts"); and they try to determine new elements that have been added. To document this, each student or group has an old map enlargement, colored markers, and a sheet of acetate or tracing paper to use as an overlay. If you have a historian accompanying you, students should be encouraged to ask questions and take notes.

3. Drawing Conclusions: Share maps and information gathered, using the ensuing discussion to revisit the last guiding questions: How do our findings help us to understand what life was like then and now? What do the changes in building use and land use suggest about what was important many years ago that is not as important today?

Materials Needed

Enlarged detail maps (copies for each student or group)
Paper
Clipboards
Acetate overlays or tracing paper
Colored pens

Assessments

Product: Students have colored and labeled the old maps clearly and accurately; the overlays are likewise complete and neat. Tool/reference point: compare with teacher's answer key

Short Answer: Students hand in journal entries about the observations they made in the field. Tool/reference point: checklist of key or main points

Extensions

Ask students to find out how their own home or dwelling sites have changed in form and use over the years. Was it the first building on the site? Sources for this information might include observation, parents, older neighbors, history books available at the public library, and refinancing or closing documents. (If you own your home or apartment, a title search was conducted as part of your closing. This document creates a chain of the different people that owned that particular property over time.) Based on the buildings and landscape of their homes and neighborhood, can students make an educated guess regarding the profession of their predecessors?

LESSON III. DRAWING THE VILLAGE AND LANDSCAPE

Focusing Question(s)

➤ What are the key components of our adopted settlement?
➤ How have these buildings changed over time? In structure? In terms of use? Why might they have changed?

Standards

Artistic Intent: Convey artistic intent from creator to viewer or listener; critique one's own and others' works in progress, both individually and in groups, to improve intent.

Sense of Place: Apply knowledge of local environment through active participation in local environmental projects.

Length of Time Needed to Complete

Minimum of 90 to 120 minutes

Procedures

1. The students, with your input and guidance, select a number of site components and/or buildings of interest and create elevation drawings and detailed drawings of these sites. Remind them of their purpose through guiding questions:

   > What were some of the community buildings on the map that are no longer present? Did we find evidence of them on site? Have any structures moved?
   > What interesting buildings/sites still remain? How has their use changed over the years?
   > Which sites/buildings would best fit into our Quest, the purpose of which is to show how the village has both changed and remained the same over the years?

2. Students, working in pairs or threesomes, select a site or building to focus on, drawing from the list in the third question above. Each group should produce one carefully detailed drawing of the "complete" site or building and at least one characteristic detail (enlargement) from that view (for example, an elevation drawing of the facade of a building and a detailed rendering of the fanlight over the doorway; or a side view of a cemetery entrance and a close-up of a single gravestone).

3. In-Class Critique:

   > Display elevation drawings and detail drawings. See if students can locate the detail drawing on the elevation drawing, connecting the two. This can be a great opportunity to discuss architectural history with students: comparing the forms of the houses and their details, teaching kids to "read" the age of a building or the presence of additions and other building modifications: new bathrooms, handicap-accessibility ramps, porches, and so forth.
   > Are the student's drawings detailed enough to be recognizable or should the artists add more precision, detail, and/or environment to

their pictures for reasons of clarity or aesthetics? These first drawings may be sketches, but they must be executed with enough detail so that a return visit to the site for additional drawing is not necessary.

➤ Another option is for students to take digital photographs as well of their buildings. This option serves two purposes: students have exposure to an emerging technology, and they have a resource they can use to continue to refine their drawings, either in-class or as homework.

Materials

Large masonite boards
Paper
Pencils
Erasers
Tape and rulers

Assessment

Product: Students have followed directions and have carefully executed, detailed drawings to show for it.

Performance: Students have offered thoughtful feedback to their classmates on their work.

Follow-up

Students should make final versions of their respective drawing(s) using black ink, which will allow the drawings to be photocopied or scanned to become part of the final Quest treasure map. (This could be homework.)

LESSON IV. PRIMARY SOURCE INVESTIGATIONS AT MUNICIPAL OFFICE, LOCAL HISTORICAL SOCIETY, OR LIBRARY

Focusing Questions

➤ Where (and from whom) can we learn more about the history of our community?

➤ How has our community changed over time? Why?

> Are there key persons, places, and events linked to these changes?

> How are these changes related to changes in regional or American life— or changes in technology?

Standards

Being a Historian: Collect and use primary resources while creating original historical interpretations.

Understanding Place: Demonstrate knowledge of past and present community heritage (e.g., traditions, livelihoods, customs, stories, changing demographics, or land use) and recognize ways in which this heritage influences their lives.

Length of Time Needed to Complete

Minimum of 120 minutes

Procedures

1. To illustrate how students can use town records for research, begin by having your students review a current or recent town report. This will help them see how much a single document can tell about the town at a particular point in time.

2. Introduce the students to the types of information housed in the local historical society, town offices, or library, depending upon which location you are using for your primary source investigations. It is important to remind students that they are in fact handling history and that it is important to follow the protocols of the hosting institution when using books and archival materials. Availability of materials and the focus of your Quest will inform how you plan this activity.

3. Research Activity: Based on the previous point—and with the permission and input of the institution—set up stations with certain resources and questions keyed to them. These resources might include old maps, town reports, cemetery records, church records, deeds, letters, photographs, early histories, or newspaper articles. Note that many historical societies have file folders containing articles, photographs, and documents regarding longstanding community families. Also, if your community has ever filed a "Historic District" application, this, too, can serve as a good resource for information regarding building history. The underlying theme of the research should be that of your Quest: changes in

structures, changes in use, changes in population, as well as the residents, activities, and businesses that flourished in that section of town. This is a good opportunity to invite community volunteers such as members of the historical society or parents to participate in the project as station monitors.

Have students, working in small groups, cycle through the stations.

After students have answered the questions at each station, they should be encouraged to come up with two additional things:

> a question of their own design, whose answer can be found using one of the source materials; and
> one other nugget of information they find interesting, perhaps because of its anecdotal quality or its relationship to the people, places, or events connected with the Quest.

4. Report out: Students share their own questions and bits of information from the preceding activity.

Materials

Primary and secondary historical sources
Pencils
Clipboards
Lined paper

Assessment

Short answer: Students hand in their answers to the target questions asked at each station.

Product: Students hand in their questions.

Performance: Students share their "nugget" or "anecdote" with the rest of the class.

Extension

Students prepare for next week's "Oral History Interviews" researching families/buildings/sites: Develop one or more research questions. Practice interviewing a family member or a friend. Refine the question and then develop a follow-up question.

LESSON V. ORAL HISTORY INTERVIEWS

Focusing Questions

➤ Who are the elders of our community?
➤ How can we learn from them?
➤ What kinds of things can we learn from them?

Standards

History, Understanding Place, Being a Historian: Use oral history methods to understand the ways in which people assign meaning to their own historical experiences.

Understanding Place: Explore the interrelationship between the local environment and local community heritage.

Length of Time Needed to Complete

Two 90-minute periods

Procedures

1. Plan the format of the oral history component. First, you need to determine just how many "sources" for oral histories you have. Understand that second- and third-generation "stories" can be just as informative as first-person accounts. Although you may be able to have children interview individuals who have links to their personal sites, if you are only able to locate a handful of subjects you might, as we did, arrange a group storytelling session in an old schoolhouse or building within your chosen settlement. In any case, you need to make sure to inform the subjects of the purpose of the session—and emphasize your Quest theme and buildings/sites along your Quest trail. Also be sure to let them know who else will be there. The camaraderie amongst the guests seems to bring out stories and anecdotes that might otherwise have been overlooked or forgotten. These sessions can provide great intergenerational photo opportunities for local media.

2. Review with your students the concept of oral history and consider its validity as a historical source, as well as its limitations, due to the "story" component and the inherent perspective of the teller. Be sure to discuss

the difference between a first-person account and a retold story, which can be embellished or romanticized with each retelling.

Students also need to understand that stories are told from the point of view of an individual. To illustrate how a point of view on the same thing can differ greatly, you might have them draw an item, such as a shell, from two very different perspectives, for example, a side view and a bird's-eye view. Even thought the object stays the same, the drawings will be very different. The interests, opinions, and point of view of the teller will likewise almost always affect the content of the oral history.

3. Prior to whatever format you use for the oral histories, students need to prepare their questions. These should be linked to the general theme of the Quest and/or the experiences of those interviewed. Students must work to develop questions that cannot be answered with a simple yes or no but instead encourage the telling of anecdotes as well as the relating of specific Quest content. Students should think of possible follow-up questions, too. As an example, ask for a volunteer to share a question. Next, have students brainstorm about how that question might be answered. Then, with the stated goals of encouraging both detailed factual answers and anecdotal responses, have the students work together to improve the stating of the question and think of additional, follow-up questions, as follows:

Student question: "Did you have to do chores before you went to school?" (generates a yes or no answer)

Change to prompt more information: "What kinds of chores did you have to do before you left for school in the morning?" (elicits a list)

Follow-up question: "Do you recall an interesting or amusing event that you—or someone you know—had happen that made getting to school on time difficult?"

4. Set students to work in their groups, getting feedback from one another about the quality and appropriateness of their questions. Encourage questions that are purely informational and those that are more general or designed to encourage anecdotal reminiscences.

5. Students do the interview or participate in the storytelling session.

6. Report out: (could be homework): If students are conducting individual interviews, they should hand in a write-up of the interview and be ready to share information or anecdotes that would be of interest to the group. If you had a group storytelling session, you might assign each student the

task of producing a written retelling of a story of interest, ideally related in some way to his or her chosen site.

Materials Needed

Clipboards
Pencils
Paper
Tape recorder (optional, but recommended)
Cameras (optional, but recommended)

Assessment

Product: Completed question(s) and follow-up questions. Short write-up summarizing answer to the question (and/or one favorite anecdote or story from the session).

Follow-up

(Could be homework.) Students generate a list of ideas regarding the information they want to include in their teaching clues.

LESSON VI. WRITING THE QUEST: CLUES THAT TEACH AND MOVE

Focusing Questions

➤ What are the key stories of this place?
➤ How can we tell them concisely—and in verse?
➤ How do we lead visitors on a playful guided tour of our adopted community?

Length of Time Needed to Complete

Two 60- to 90-minute periods

Standards

Writing Dimensions: Draft, revise, edit, and critique written products so that final drafts are appropriate in terms of the following: purpose, organization, details, and voice or tone.

Continuity and Change: demonstrate an understanding that perceptions of change are based on personal experiences, historical and social conditions, and the implications of change for the future.

Understanding Place: Explore the interrelationship between the local environment and local community heritage

Procedures

1. Before they start trying to write their clues, it is important to make your expectations very clear to the students and to reestablish the purpose or intent of the Quest in order to focus their writing.
2. Lead a discussion using the following guiding questions:

   ➤ What are the unique or interesting visual and historical elements that you might want to include in your clues? Have students share from the lists that have already been made.
   ➤ Where should the Quest begin and end? Work as a class to establish the route and the order of sites to be visited. Recall the attributes of good starting places (easy to find, parking, safe) and good hiding spots (secure, off the beaten path).

3. Review clues that move you through space and the clues that teach.
4. Distinguish between a good clue, a better clue, and the best clue. A "good" clue will work to move a visitor in the right direction but not engage the sensory experience of the surroundings, for example, "Go twenty paces and turn right." A "better" clue might say, "With a flashing light in sight turn right." Here, the clue picks up a feature, the flashing light, from the environment. The "best" clue might read, "The ol' Vermont sugar maker says, 'Wind from the east, sap runs the least. Wind from the west, sap runs the best.' It's a bad sap day so turn that way."
5. Using photocopies of the student's drawings, lay out a map of your village on a large table, a bulletin board, or the floor (the floor works best,

because it allows students to walk through the space). Add roads, trails, and watercourses using masking tape and yarn. Use string or scotch tape to lay out the proposed route of the Quest. Discuss with the students the need for both clues that teach about the buildings or sites and clues that will move the Questgoers for one site to the next.

6. If you have the luxury of time, break students up into pairs and have them take turns guiding one another to different places in the school or on the playground by various moving clues: body orientations of left and right, directions of north, south, east, and west, pacing, compass reading, etc. When they reconvene, ask what kind of directions they used and which worked the best. It is helpful to let them discuss the strengths of weaknesses of each type of clue.

7. Working with the students, designate to each group the responsibility for producing a specific "moving" clue as well as a "teaching" clue. Group 1 is responsible for covering the ground from point A to point B, while teaching about site 1; Group 2 is responsible for covering the ground from point B to point C, while teaching about site 2; and so on.

8. Activity: Inform the students of the formatting requirements for their clues. In the example that follows, each of the students was asked to produce at least two couplets having nine to twelve beats in each line: "On the old stone foundation, beside the stream,/There once stood a mill, powered by water, not steam." Encourage them to make lists of rhymes for some of their key words. Then set them to work. Expect some struggling, but once they start coming up with doggerel verse, it becomes catchy. Remind them that their first effort may not be their best effort, and that they should work to refine their words so that the clues teach clearly, include interesting information, and transition smoothly from one place to another.

9. Report out: Volunteers read what they have so far and give and receive peer/teacher feedback.

Materials

Copies of student drawing
Tape
Yarn
String
Pencils
Paper

Assessment

Product: Students/groups have produced at least a draft of one or two appropriate couplets, either a clue that teaches, one that moves, or both.

Follow-up

Students finish their work, focusing on accuracy and clarity of facts, descriptive language, and imagery. (This could be homework.) Using the drafts of the clues, walk through the Quest to see if it works. Afterward, students should debrief: How did it go? Where do we need more clarity? Was there enough teaching information? With this feedback in mind, student should refine and improve their clues to a "final" state.

LESSON VII. PULLING IT ALL TOGETHER

Focusing Questions

What do we still need to do in order to have: A finished map? Polished clues? A completed treasure box?

Standards

Design and Production: Design and create media products that successfully communicate.

Understanding Place: Explore and participate in sustaining or building on unique and valued elements of past and present community heritage.

Length of Time Needed to Complete

One to two 90-minute classes. Note: This can also be done as project homework. Depending on how much they have done already, what state your Quest is in, and whether this is going to be in-class work or homework, you may choose to give this lesson more or less time.

Procedure

Show a sample of a finished Quest to students and encourage discussion about design, layout, graphics, and so on. Tell them that this is the time for them to

Quest	Exemplary	Commend-able	Adequate	Weak	Unacceptable
Teamwork	Also, showed leadership throughout project	Also, exceeded expectations and helped others do so as well	Did share of work AND met expectations in terms of content quality and timeliness	Did share of work OR met expectations in term of content, quality and timeliness	Neither did share of work nor met expectations in terms of content, quality and timeliness
Research	Also, did all extension activities and extras	Also, did some extension activities and extras	Research activities accurate and complete	Research activities are accurate OR complete	Research activities show little effort and are neither accurate nor complete
Drawing	Also, includes setting elements; a striking likeness	Site drawing well-detailed, clear and accurate	Site drawing clear and accurate	Site drawing either clear or accurate	Site drawing sloppy and lacks care
Clues	Also, uses vivid descriptors, good historical vocabulary, and excellent meter and flow	Also, uses vivid descriptors or good historical vocabulary or has excellent meter and flow	Writing is understandable, includes moving or teaching component and shows proper grammar usage and mechanics	Writing lacks, or does not include a moving or teaching component, or shows proper grammar usage and mechanics	Writing lacks clarity, does not include a moving or teaching component and neglects standard writing conventions

A.2 This could be used as an ongoing assessment, filling out the rubric after students complete each component.

pull together and work to bring their Quest to a final product. Set a deadline for the completion of tasks. Develop small teams, with each team taking on one or more of the tasks listed below:

➤ Cartographers: Photocopy student drawings, reduce the copies, and then place them all together onto a single map.
➤ Artists: Design the compass rose, indicating north. The best compass rose will be based on some inherent quality of your Quest site, for example, a detail of a building or a distinctive weathervane.
➤ Editors: Type, proofread, and align the clues into a single text.

The ANNOTATED Beaver Meadow Quest

Student written clues:	_Commentary:_
Down a hill, to a white steeple, Long ago, it gathered people;	_Movement clue_
In this chapel - what is it they do? Pray, play cards and sell stuff not new.	_Changing use – thrift sales_
Up the steps, feet on porch floor, Look for the star to the left of the door.	_Noticing detail – tied to student Drawing of flagpole_
Walk south-east away from the chapel, Where once a house, now grow trees of apple,	_Movement clue Changing use – once cider mill, now a small apple orchard_
A mill stood here where cider was pressed, Walk up to the corner, then take a rest,	_Movement clue_
Turn left; third house on the north-west side, Here once the first school house did reside.	_Change: 1st school house, gone now a private home_
After chores likes milking the cow, Kids walked to school, unlike now.	_Clue from Oral History_
On your right is the school were you might have learned, Built in 1922, when the first one burned.	_Change—Fire –clue from historical Society. 2nd Beaver Meadows School House_

A.3 The annotated Beaver Meadow Quest illustrates how student learning links back to the standards, and how their writing reflects and incorporates the various lessons and activities.

- Designers: Organize and lay out all of the Quest components into a single publication.
- Boxmakers: Make and/or decorate the treasure box.
- Box Hiders: Locate an appropriate location at the end site for hiding the box and create any needed modifications, such as installing a false birdhouse or a hidden shelf.
- Bookmakers: Make a sign-in book and information booklet for the treasure box. The information booklet is composed of student essays and other leftover material: photocopies of the map, photographs, drawings, and journal entries.
- Stampmakers: Design and carve a unique stamp.
- Poets: Make up introductory, closing, and fill-in rhymes, and ensure that clues hook together nicely.

Materials

Paper
Pencils
Glue sticks
Computers
A pair of compasses (artist)
Various art materials (boxmaker, bookmaker, and stampmaker)

Assessment

Product: Students have completed their part of the Quest. Tool/reference point: task-specific rubric attached

LESSON VIII. TESTING YOUR QUEST

Once you have a complete treasure box, a map, and a polished set of clues, you are ready for testing. Get a volunteer—another class, a group of parents, or a group of seniors—to "test drive" your Quest and make sure it works. After testing, incorporate all recommended changes and then make your final draft as the final lesson. When you finish, throw a party to acknowledge your students' hard work and generous offering to the community!

Taking it public

Hold a public event to share the Quest with the broader community. Consider aligning this with another already planned community celebration. Be sure to send every student home with at least one copy of the completed Quest. Also remember to submit the completed Quest to your local paper and to your local Quest program organizers.

NOTES

Introduction: Place-Based Education (pp. 1–4)

1. Quoted in Jack Chin, "Connecting Schools and Communities through Place-Based Education," a background paper for the Funders' Forum on Environment and Education (F2E2) Conference held April 11–12, 2001, 17.

1. The Joy of Treasure Hunts (pp. 5–13)

1. Annie Dillard, *Pilgrim at Tinker Creek* (New York: Harper and Row, 1974) 14.
2. Daniel Jantos, interview by Delia Clark, March 2002, Woodstock, Vt.
3. Quoted in *Tracking the Dragon* (Chimacum, Wa.: Wild Olympic Salmon, 1991), back cover.

2. The Story of Questing (pp. 14–26)

1. John Elder, quoted in Steve Glazer, ed., *Valley Quest: 89 Treasure Hunts in the Upper Valley* (White River Junction, Vt.: Vital Communities, 2001), inside cover.
2. Anne Swinscow and John Hayward, *Dartmoor Letterboxes* (Devon, U.K.: Kirkford Publications, 1988), 45.
3. Ibid., 65.

3. The Spirit of a Place (pp. 27–44)

1. We are indebted to Carola Lea, the Lyme historian, and the Lyme Community Elders for their assistance in creating the Lyme Center Dances Quest.

5. Varieties of Quests (pp. 61–79)

1. We are indebted to David Sobel for his work on the original version of the Quest Taxonomy, which he developed as part of a Quest workshop.

6. Notes for Teachers (pp. 80–98)

1. Maggie Stier, *Valley Quest Teachers Guide* (White River Jct, Vt.: Vital Communities, 1998), 10–11.

7. Out in the Field (pp. 99–113)

1. Walt Whitman, *Leaves of Grass* (Garden City, N.J.: Doubleday, 1926), 26.
2. Philip Booth, "How to See Deer," in *Relations: New and Selected Poems* (New York: Viking, 1975), xx.
3. Many thanks to Ted Levin for his hand in creating the Miraculous Tree Quest.
4. Ted Levin, *Blood Brook: A Naturalist's Home Ground* (White River Junction, Vt.: Chelsea Green, 1992), 117–119.

8. Researching Your Place (pp. 114–136)

1. Thanks to the Vermont Folklife Center for developing and sharing these guidelines.
2. Monica Edinger, *Seeking History: Teaching with Primary Sources in Grades 4–6* (Portsmouth, N.H.: Heinemann, 2000), viii.
3. Arthur Sze, "The Network," taken from *The Redshifting Web: Poems 1970–1998* (Port Townsend, Wa.: Copper Canyon Press, 1998), 122.
4. We are indebted to Lois Overton and Charles Latham for their research, editing, and assistance in creating the George Knox Quest.

10. Making Maps (pp. 155–171)

1. Christopher Alexander, *The Timeless Way of Building* (New York: Oxford University Press, 1979), 101.

13. Building a Quest Program (pp. 203–212)

1. Dick Norton, M.D., phone interview and email correspondence with Delia Clark, April 25, 2002.
2. Dave Monk, "Application for Assistance with Quest Development," submitted to the National Park Service Rivers and Trails Program, 1999.

14. From Questing to Stewardship (pp. 213–216)

1. Wendell Berry, *What Are People For?* (New York: North Point, Farrar, Straus, and Giroux, 1988), 169.

BIBLIOGRAPHY

Alexander, Christopher, et al. 1977. *A Pattern Language.* New York: Oxford University Press.

Baldwin, Jessie A. 1983. *History and Folklore of Post Mills, Vermont.* Thetford, Vt.: Thetford Historical Society.

Bateson, Mary Catherine. 1994. *Peripheral Visions: Learning Along the Way.* New York: HarperCollins.

Baylor, Byrd. 1974. *Everybody Needs a Rock.* New York: Aladdin.

———. 1991. *Your Own Best Secret Place.* New York: Atheneum.

Beatley, Timothy, and Kristy Manning. 1997. *The Ecology of Place.* Washington, D.C.: Island Press.

Berry, Wendell. 1977. *The Unsettling of America.* New York: Avon.

Blouin, Glen. 2001. *An Eclectic Guide to Trees.* Erin, Ontario: Boston Mills Press.

Caduto, Michael J., and Joseph Bruchac. 1994. *Keepers of Life.* Golden, Colo.: Fulcrum.

———. 1997. *Keepers of the Animals.* Golden, Colo.: Fulcrum.

Carroll, David M. 1999. *Swampwalker's Journal.* Boston: Houghton Mifflin.

Carse, James P. 1994. *Breakfast at the Victory: The Mysticism of Ordinary Experience.* San Francisco: HarperSanFrancisco.

Center for Ecoliteracy. 2000. *Ecoliteracy: Mapping the Terrain.* Berkeley, Calif.: Center for Ecoliteracy.

Cobb, Edith. 1993. *The Ecology of Imagination in Childhood.* Dallas: Spring Publications.

Cole, Luane, ed. 1976. *Patterns and Pieces.* Canaan, N.H.: Phoenix Publishing.

Conant, Roger, and Joseph T. Collins. *Reptiles and Amphibians.* Boston: Houghton Mifflin.

Connor, Sheila. 1994. *New England Natives: A Celebration of People and Trees.* Cambridge, Mass.: Harvard University Press.

Daubenspeck, Mary, and Judith G. Russell, eds. 2000. *We Had Each Other: A Spoken History of Lyme, New Hampshire.* Lyme, N.H.: The Lyme Oral History Project.

Deetz, James. 1977. *In Small Things Forgotten: The Archeology of Early American Life.* New York: Anchor Books.

DeGraff, Richard M., and Mariko Yamasaki. 2001. *New England Wildlife.* Hanover, N.H.: University Press of New England.

Dillard, Annie. 1974. *Pilgrim at Tinker Creek.* New York: Harper and Row.

Dragonwagon, Crescent. 1990. *Home Place.* New York: Aladdin.

Edinger, Monica. 2000. *Seeking History: Teaching with Primary Sources in Grades 4–6.* Portsmouth, N.H.: Heinemann.

Elder, John. 1998. *Reading the Mountains of Home.* Cambridge, Mass.: Harvard University Press.

Elder, John, ed. 1998. *Stories in the Land: A Place-Based Environmental Education Anthology.* Great Barrington, Mass.: The Orion Society.

Foster, David R., and John F. O'Keefe. 2000. *New England Forests Through Time: Insights from the Harvard Forest Dioramas.* Cambridge, Mass.: Harvard University Press.

Garvin, James L. 2001. *A Building History of Northern New England.* Hanover, N.H.: University Press of New England.

Glazer, Steven, ed. 1999. *The Heart of Learning: Spirituality in Education.* New York: Jeremy P. Tarcher.

―――. 2001. *Valley Quest: 89 Treasure Hunts in the Upper Valley.* White River Junction, Vt.: Vital Communities.

Goldsmith, Edward. 1993. *The Way: An Ecological World-view.* Boston: Shambhala.

Hale, Jonathan. 1994. *The Old Way of Seeing.* Boston: Houghton Mifflin.

Hannum, Hildegarde. 1997. *People, Land and Community.* New Haven: Yale University Press.

Hayward, Jeremy. 1997. *Letters to Vanessa on Love, Science and Awareness in an Enchanted World.* Boston: Shambhala.

Hayward, John. 1992. *101 Dartmoor Letterboxes.* Devon, U.K.: Kirkford Publications.

Heinrich, Bernd. 1997. *The Trees in My Forest.* New York: HarperCollins.

Henderson, Robert K. 2000. *The Neighborhood Forager.* White River Junction, Vt.: Chelsea Green.

Hinchman, Hannah. 1997. *A Trail Through Leaves: The Journal as a Path to Place.* New York: W. W. Norton.

Hiss, Tony. 1990. *The Experience of Place.* New York: Alfred A. Knopf.

Hubka, Thomas. 1984. *Big House, Little House, Back House, Barn.* Hanover, N.H.: University Press of New England.

Hughes, Charles. *The Mills and Villages of Thetford, Vermont.* Thetford, Vt.: Thetford Historical Society.

Jackson, Wes. 1994. *Becoming Native to This Place.* Washington, D.C.: Counterpoint.

Johnson, Charles. 1998. *The Nature of Vermont.* Hanover, N.H.: University Press of New England.

Kohl, Judith and Herbert. 2000. *The View from the Oak: The Private Worlds of Other Creatures.* New York: The New Press.

Kricher, John, and Gordon Morrison. 1988. *A Field Guide to Eastern Forests.* Boston: Houghton Mifflin.

Kunstler, James Howard. 1993. *The Geography of Nowhere.* New York: Simon and Schuster.

Latham, Charles. 1972. *A Short History of Thetford, Vermont.* Thetford, Vt.: Thetford Historical Society.

Lawrence, Gail. 1998. *A Field Guide to the Familiar.* Hanover, N.H.: University Press of New England.

Leopold, Aldo. 1949. *A Sand County Almanac.* Oxford: Oxford University Press.

Levin, Ted. 1987. *Backtracking: The Way of a Naturalist.* White River Junction, Vt.: Chelsea Green.

―――. 1992. *Blood Brook: A Naturalist's Home Ground.* White River Junction, Vt.: Chelsea Green.

Lingelbach, Jenepher, ed. 1986. *Hands-On Nature: Information and Activities for Exploring the Environment with Children.* Woodstock, Vt.: VINS.

Ludwig, Allan I. 1966. *Graven Images: New England Stonecarving and Its Symbols, 1650–1815.* Hanover, N.H.: University Press of New England.

MacMahon, James A. 1985. *Deserts.* New York: Alfred A. Knopf.

Mander, Jerry. 1991. *In the Absence of the Sacred.* San Francisco: Sierra Club.

Mansfield, Howard. 1993. *In the Memory House*. Golden, Colo.: Fulcrum.

McNamee, Gregory. 1993. *Named in Stone and Sky: An Arizona Anthology*. Tucson: University of Arizona Press.

Merrill, Christopher. 1991. *The Forgotten Language*. Salt Lake City: Gibbs-Smith.

Muir, Diana. 2000. *Reflections in Bullough's Pond: Economy and Ecosystem in New England*. Hanover, N.H.: University Press of New England.

Nabhan, Gary Paul. 1997. *Cultures of Habitat*. Washington, D.C.: Counterpoint.

Nabhan, Gary Paul, and Stephen Trimble. 1994. *The Geography of Childhood*. Boston: Beacon.

Orr, David W. 1994. *Earth in Mind: On Education, Environment and the Human Prospect*. Washington, D.C.: Island Press.

Osborne, Terry. 2001. *Sightlines: The View of a Valley Through the Voice of Depression*. Hanover, N.H.: Middlebury College Press/University Press of New England.

Pakenham, Thomas. 1996. *Meetings with Remarkable Trees*. New York: Random House.

Palmer, Parker. 1983. *To Know As We Are Known*. San Francisco: HarperSanFrancisco.

———. 1997. *The Courage to Teach*. San Francisco: Jossey-Bass.

Parrella, Deborah. 1995. *Project Seasons: Hands-on Activities for Discovering the Wonders of the World*. Shelburne, Vt.: Shelburne Farms.

Rezendes, Paul. 1999. *Tracking and the Art of Seeing*. New York: HarperCollins.

Rockefeller, Steven C., and John C. Elder. 1992. *Spirit and Nature: Why the Environment Is a Religious Issue*. Boston: Beacon.

Sale, Kirpatrick. 1985. *Dwellers in the Land: The Bioregional Vision*. San Francisco: Sierra Club.

Sanford, Rob, and Don and Nina Huffer. 1995. *Stonewalls and Cellarholes: A Guide for Landowners on Historic Features and Landscapes in Vermont's Forests*. Waterbury, Vt.: Vermont Agency of Natural Resources.

Sauer, Peter, ed. 1992. *Finding Home: Writing on Nature and Culture from Orion Magazine*. Boston: Beacon.

Shepard, Paul. 1982. *Nature and Madness*. San Francisco: Sierra Club.

Sloan, Eric. 1967. *An Age of Barns*. New York: Ballantine.

———. 1972. *The Little Red Schoolhouse*. New York: Doubleday.

Snyder, Gary. 1980. *The Real Work: Interviews and Talks, 1964–1979*. New York: New Directions.

———. 1990. *The Practice of the Wild*. San Francisco: North Point Press.

———. 1995. *A Place in Space*. Washington, D.C.: Counterpoint.

Sobel, David. 1993. *Children's Special Places*. Tucson, Ariz.: Zephyr Press.

———. 1998. *Mapmaking with Children: Sense of Place Education for the Elementary Years*. Portsmouth, N.H.: Heinemann.

———. 2004. *Place-Based Education: Connecting Classrooms and Communities*. Great Barrington, Mass.: The Orion Society.

Stilgoe, John R. 1998. *Outside Lies Magic: Regaining History and Awareness in Everyday Places*. New York: Walker.

Swinscow, Anne. *Dartmoor Letterboxes*. 1984. Devon, U.K.: Kirkford Publications.

Thomashow, Mitchell. 1995. *Ecological Identity: Becoming a Reflective Environmentalist*. Cambridge, Mass.: MIT Press.

Tuxill, Jacquelyn. 2000. *The Landscape of Conservation Stewardship*. Woodstock, Vt.: Marsh-Billings-Rockefeller National Historic Park.

Visser, Thomas Durant. 1997. *Field Guide to New England Barns and Farm Buildings*. Hanover, N.H.: University Press of New England.

Vitek, William, and Wes Jackson. 1996. *Rooted in the Land*. New Haven: Yale University Press.

Wessels, Tom. 1997. *Reading the Forested Landscape: A Natural History of New England*. Woodstock, Vt.: Countryman Press.

White, Gilbert. 1986. *A Selbourne Year: The Naturalist's Journal for 1784*. Devon, U.K.: Webb and Bower.

ACKNOWLEDGMENTS

Together, Delia and Steve extend their gratitude to:

Betty Porter for her vision, her early and on-going commitment to Questing, and her friendship.

David Sobel, Maggie Stier, Linny Levin, Ginger Wallis, Uri Harel, and Sue Kirincich, who contributed so much to the evolution of Questing.

Ted Levin, Maggie Stier, Simon Brooks, Andy Boyce, Ros Seidel, Tim Traver, Heewon Shin, Becky French, and Stacey Glazer for their thoughtful and timely reading of our manuscript.

Simon Brooks, Jon Gilbert Fox, Ted Levin, and Jim Sheridan for their beautiful photographs.

Michael Duffin for his First Class assistance.

The Robins Foundation, Anne Slade Frey Charitable Trust, Ellis L. Phillips Foundation, and Upper Valley Community Foundation for their early support of Valley Quest.

In addition, Delia offers her heartfelt thanks to:

Tim, Kalmia, Mollie, and Toben Traver—my lifelong soulmates in the big Quest.

Susan Clark and Mark Bushnell for unflagging, good-humored confidence and support, including the offer of a camp stocked with loving incentives.

Harry Clark for taking me on my first mystery rides and Virginia Clark for teaching me to write.

Steve Glazer, David Sobel, and Bo Hoppin for making collegiality fun.

Derry Tanner and the Sudbury Foundation, whose financial support and confidence in me made my participation in this book possible.

Doug Evans and Charlie Tracy of the National Park Service Rivers and Trails Program, Dave Monk from the Salt Ponds Coalition, and Robin Jorgensen and Wendy Siden from the Beebe Environmental and Health Sciences Magnet School, for lending their energy and enthusiasm to Questing.

Daniel Jantos for sharing stories of youthful treasure hunts.

Kevin Dann for offering me seclusion in the surreal serenity of Riverlee, and Susan Inui-Kelley for giving me the proverbial "white room" when nothing else would do.

Megan Camp, Jessica Brown, and Nora Mitchell, for including Questing in their grand vision.

Dick Norton for thoughtful reflections about how to make a Quest program sustainable.

Katka Rajcova for translating my favorite Quest into Slovak for Babynec.

John Souter and Silvia Provost for their initial leap of faith in bringing Questing into Mollie's fourth grade classroom, and for their ongoing contributions as advisors.

Paul Bocko, David Clark, Aanika DeVries, Michael Duffin, Jim Gruber, Sandy Hamon, Alan McIntyre, Wendy Oellers, Jen Risley, and Stephanie Tuxill, for their support of Questing and this book.

The albino deer on Cumberland Island and the flying squirrel on Happy Valley Road.

Steve bows with gratitude to:

My family for giving me time, space, and the inclination to explore.

Rhoda Radow, a high school teacher who led me out of the classroom and into the world.

Professor Frank Griggs, whose course in Industrial Archeology in Bath, England, brought history to life.

Stacey Glazer, my best friend and traveling companion since 1984.

Sheila Hixon for love, support, and friendship along the path.

Chagdud Rinpoche for teaching me to appreciate, give thanks, and pay attention.

My Living Education friends for helping me to discover place.

Delia, the staff, and the board of Vital Communities, for wooing me "east" and "home."

Bill Aldrich, Marguerite Ames, Charles Balch, Susan Bonthron, Andy Boyce, Len Cadwallader, Monique Cleland, Don Cooke, Steve Dayno, Ken Elder, A Forest for Every Classroom, Becky French, Sara Goodman, Lisa Johnson, Amos Kornfeld, Charles Latham, Carola Lea, Robin Leonard, Linny Levin, Ted Levin, Promise of Place, Betty Porter, Sylvia Provost, Dorf Sears, Ros Seidel, Binx Selby, Bill Shepard, Jim Sheridan, Heewon Shin, Anne Silberfarb, Fern Tavalin, Heather Trillium, Beckley Wooster, Diana Wright, the Anne Slade Frey Charitable Trust, The Boston Foundation, The Bay Paul Foundation, Connecticut River Joint Commissions, the Lyme Foundation, the

Ellis L. Phillips Foundation, the Spirit in Community Fund, the Upper Valley Community Foundation, the Walker Fund of New Hampshire Charitable Foundation, and the Wellborn Ecology Fund of the Upper Valley Community Foundation, for help, friendship, and the support of Valley Quest.

The A. D. Henderson Foundation, the Alexander and Adelaide Hixon Foundation for Religion and Education, the Robins Foundation, and the Windham Foundation, for their generous support of the writing of this book.

Beckley, Gabe, Len, Lisa, and Stacey. "Let's be careful out there!"

Simon Brooks, for friendship, support, editing, photography, stamp cutting, and helping give birth to this book.

Stacey, Kayla Beth, and Emma Rose Glazer, my partners on the path. I love you.

And finally, we both give thanks for the gift of this moment and teachings of this place.

INDEX

ABOUT THE AUTHORS

Delia Clark co-founded Antioch New England Institute as a community-service extension of Antioch New England Graduate School and has served there as Co-Director and Project Director. While there she co-founded Vital Communities of the Upper Valley and served as Executive Director, helping to launch Valley Quest. Delia's current work focuses on place-based education, civic engagement, and facilitating community dialogue. She is a frequent trainer and speaker in these areas nationally and internationally for organizations including the National Park Service Conservation Study Institute, The Conservation Fund, QLF Atlantic Center for the Environment, and Shelburne Farms. She is a recipient of the New England Environmental Education Alliance award for Outstanding Contributions to Environmental Education. Delia is a graduate of the University of Vermont and Antioch.

Steve Glazer coordinates the Valley Quest program for Vital Communities, a regional nonprofit organization serving Vermont and New Hampshire. Steve has pioneered place-based education—the use of the local community as the integrating context for curricular studies—in his work with the Patagonia, Arizona, public schools, dozens of schools across New England, and the Valley Quest program. Cofounder of The Naropa Institute School of Continuing Education and Living Education: Place-based Living and Learning, Steve is the author of *The Heart of Learning* (Tarcher/Putnam, 1999), and editor of *Valley Quest: 89 Treasure Hunts in the Upper Valley* (Vital Communities, 2001), and *Valley Quest II: 75 More Treasure Hunts in the Upper Valley* (Vital Communities, 2004). He is a graduate of Union College and the University of Chicago.

Proceeds from the sale of this book will support place-based educational programming at Antioch New England Institute and Vital Communities.

Antioch New England Institute promotes a vibrant and sustainable environment, economy, and society through informed civic engagement. A nonprofit environmental and educational consulting organization of Antioch New England Graduate School, the institute provides training and resources to communities and organizations in the following areas: environmental education, leadership training, environmental policy development, nonprofit management and governance, facilitation, and democracy building.

Vital Communities works to engage citizens in community life and to foster the long-term balance of cultural, economic, environmental, and social wellbeing in the Upper Valley region of Vermont and New Hampshire. Vital Communities has three core program areas: Valley Quest's educational treasure hunts celebrate our community's special places; Valley VitalSigns draws people together to consider and improve the quality of life for all in our region; and Community Profiles are facilitated single or multitown forums where residents come together to find solutions to common concerns. Web site: www.vital-communities.org.